OPERATIONAL PROCEDURES DESCRIBING PHYSICAL SYSTEMS

OPERATIONAL PROCEDURES DESCRIBING PHYSICAL SYSTEMS

Maricel Agop
Ioan Merches

CISP

CRC Press
Taylor & Francis Group
Boca Raton London New York

CRC Press is an imprint of the
Taylor & Francis Group, an **informa** business

CRC Press
Taylor & Francis Group
6000 Broken Sound Parkway NW, Suite 300
Boca Raton, FL 33487-2742

Printed on acid-free paper

International Standard Book Number-13: 978-0-367-02440-6 (Hardback)

Visit the Taylor & Francis Web site at
http://www.taylorandfrancis.com

and the CRC Press Web site at
http://www.crcpress.com

"*Intelligence is the ability to adapt to change*"
Stephen Hawking

Contents

Preface

The special theory of relativity can be simply taken as the realization of the fact that, no matter if the distance and the length physically mean the same quantity they have the same units from the point of view of measurement they are essentially different. The difference is clearly apparent in the way they are measured and is always explicit in the imaginary experiments derived from practical knowledge based on the conclusion of which Einstein founded the theory. Indeed, the distance refers to space, and is measured with light signals, which entails convention on the simultaneity of the events regarding the light. On the other hand, the length is a characteristic of the matter and is measured by comparison with a standard length, for instance a rule or a tape.

Apparently, the measurements of length do not involve physical conventions, akin to those associated with the simultaneity. The only kind of conventions of a theoretical nature here seem to be purely mathematical, associated with the properties of the numbers representing the values of the length rational, real, archimedean, nonarchimedean, and so on. Even here, though, we are not free from some physical conventions, however of an entirely different kind. For instance, for the formal justification of the general theory of relativity, Einstein shows the necessity of a non-Euclidean geometry, based on length measurements along a circumference in rotation, If such a geometry is a metric one, then the representation of matter in the manner of Einstein i.e. as determining of the metric tensor in a precise geometrical way involving the curvature of the metric falls into contradiction with the geometry (see De Sitter Einstein debate in

The collected papers of Albert Einstein, Princeton University Press; the relevant papers are presented in Volume 8, Part A). The essence of this contradiction is simple: any metric of the kind assumed by Einstein can indeed be assumed a priory, but this does not mean that it necessarily represents matter. No wonder then, one comes again and again upon the fact that the distance and the length are different, even though this statement is never made explicit (for instance, the explanation of the Olbers paradox), or one finds that the Universe is necessarily closed, from a geometrical point of view (Einstein, 1916), so that the distance too not only the length from which Einstein started building his theory could be also measured on the brim of the Universe.

The fact of the matter here is to decide if Einstein was right indeed in associating a metric with the space measurements in general, and only then analyze if a kinematic cosmological model fits into this scheme. From this point of view, the space measurements no matter if length distance or something else can be framed into an all-inclusive scheme of quantum measurement, whereby the representation of a 'space quantity' is given by 2x2 matrix with two real and distinct eigenvalues, representing that quantity in its two instances: distance and length proper, as it were in the case of 'space quantity'. This quantal scheme is customary in the theory of isospin (Leon Rosenfeld, Nuclear forces, Amsterdam, 1948, Chapter IV), or even in the theory of elementary particle spin (Matthew D. Schwartz, An Axiomatic Deduction of the Pauli Spinor Theory, Int J Theor Phys, 1977), however it needs an amendment in order to describe the case in point here. Provided such an amendment, it leads with mathematical inevitability to the idea of metric, as used by Einstein and as, in fact, it is currently used within theory of relativity having thereby, however, a precise mathematical meaning, which can afterwords be translated into a physical meaning, based on the principles of the two new mechanics of the last century, quantum and wave mechanics (and this analysis could be continued...).

All of these show, as Penrose emphasized (Roger Penrose, The Road To Reality: A Complete Guide to the Laws of the Universe, Jonathan Cape, London, 2004), that the physics and in particular the General Relativity are actually far from the basic "Einsteins doctrine". Especially come to attention those theories based on the idea of an operational metric definition, such as Milne's Kinematic Relativity (Milne, E. A., Oxford, 1948). But neither Milne, nor someone else, could give an operational definition of coordinates by means of the usual velocity of signal propagation, according to which both synchronization condition and spatial distance definition are working hypotheses.

The main idea of this book has emerged from the authors' belief that, based on operational procedures offered by the measurement method analysis, some principles of modern physics apparently invi dependent from each other are, in fact, intrinsically dependent. To this end we shall introduce, as shall be seen in the following investigation, a unitary procedure of description of physical system dynamics based on Skyrmes theory (T. H. R. Skyrme, A Non-Linear Field Theory, Proceedings of the Royal Society, Vol. A 260, 1961) as the only theory naturally suited for explanation of motion of matter structure from small to large scales (Selected Papers, with Commentary, of Tony Hilton Royle Skyrme, World Scientific Series in 20th Century Physics, Vol. 3, Edited by: G E Brown, State University of New York, 1994).

This book is structured on six chapters, as follows.

The classical treatment of Keplers problem, as exposed in Chapter I, leaves room for the description of the space region of the central body by a hyperbolic geometry. If the correspondence between the empty space and the space filled with matter is taken to be a harmonic mapping, then the region of atomic nucleus, like the one of the Sun for the proper planetary system, is described by hyperbolic skyrmions. This fact makes it possible the description of the nuclear matter within the framework of general relativity. The classical "hedgehog" solution for skyrmions can then be classically interpreted in terms of the characterization of intra-nuclear forces.

There are two basic principles of the General Relativity. The first concerns determinability of the space-time metric from the matter energy- momentum tensor by means of Einsteins field equations which, in their turn, are equivalent to the principle of identity of gravitational and in- ertial mass. The second refers to the bodies motion and states that this is performed on the space-time geodesics which, this way, stand for legitimate representative of the gravitational field.

But none of these principles has been used, as such, during the time. The Einstein field equations provide non-Euclidean metrics even when the energy-momentum tensor is null. In this respect, it is nonnatural to a priory presume an energy tensor; it has to be calculated. as soon as the field equations have been solved.

On the other hand, the geodesic principle of motion obviously faces the metric ignorance. In some special cases of employed coordinates (Schwarzschild, Fock, Kruskal, etc.) adequate metrics can be obtained, while the geodesic principle almost always leads to a Keplerian problem which, as we have proved in Chapter II, is exactly solvable by means of elliptic functions.

The determinability of the vacuum or electromagnetic vacuum space-time metric, as shown in Chapter III, obeys a variational principle capable to determine such a metric by studying the motion on torsion-free trajectories only, which naturally generalize the classical uniformly accelerated motion. The group of transformations laying at the basis of this variational principle proves to be essential in regard to description of "propagation" - or, equivalently - of synchronization in a physical frame. As an example, it facilitates obtaining a clock "phase" distribution in a frame, as an invariant function: obviously, its about the analogous to de Broglies wave, lying at the basis of the well-known principle of wave particle duality. The same variational principle proves to be essential in the newest "edifices" of theoretical physics: supergravity and gauge theories.

But the most important conclusion of our analysis, described in Chapter IV, concerns the gravitational field itself. Its manifestation reflects in a continuous deformability of body systems or, more abstractly, of material points. Therefore, a stationary gravitational field can exist only for isolated bodies. Describing this deformability - as it usually does - by a deformation matrix, the space time metric can be obtained by means of the already mentioned variational principle. This way, one straightly passes from appearance (deformability of systems of material points) to essence (gravitational field), without any supplementary hypotheses on stationarity, isolation. etc., dedicated to facilitate metric obtaining, while the variational principle can prove its utility even for non-stationary, general fields.

The matter, being extended in space, should be first characterized by a surface of separation from the empty space. This surface cannot be neatly, i.e. purely geometrically defined. When it comes to extended particles, which thereby are to be considered the fundamental structural units of the matter, the physical evidence points out that they are not even stable: they are in a continuous transformation, and so is their limit of separation from space. Chapter V describes a concept of extended particle with special emphasis on this limit of separation. It turns out that the properties of inertia, as classically understood, are intrinsically related to the spin properties of quantum origin. Thus, an extended particle model cannot be but “holographic” when it comes to embedding it in a physical structure. The spin properties turn out to be essential, inasmuch as they decide the forces of interaction issuing from particles.

As presented in the last chapter, the color is an interaction property of the interaction of light with matter. Classically speaking it is therefore akin to the forces. But while forces engender the mechanical view of the world, the colors generate the optical view. One of the modern concepts of interaction between the fundamental particles of matter the quantum chromodynamics aims to fill the gap between mechanics and optics, in a specific description of strong interactions. We show here that this modern description of the particle interactions has ties with both the classical and quantum theories of light, regardless of the connection between forces and colors. In a word, the light is a universal model in the description of matter. The description involves classical Yang-Mills fields related to color. All these considerations are detailed by several simple examples in the present monograph, which is dedicated to master and PhD students, and to researchers in the field of physics as well.

Iaşi, October, 2017

Maricel Agop
Ioan Mercheş

CHAPTER 1

From Kepler problem to skyrmions

1.1 Classical Kepler motion: The position of central body

The classical Kepler motion can be described with the Newtonian equations of motion

$$\ddot{\boldsymbol{r}} + \frac{K}{r^2}\frac{\boldsymbol{r}}{r} = \boldsymbol{0}. \tag{1.1}$$

Here, K is a constant, $\boldsymbol{r}$ denotes the position vector of the material point whose motion is described, with respect to the center of force, and a dot over a symbol means derivative with respect to time. The constant K does not depend on quantities related to the point in motion, but only in cases when electric forces are involved. We can simplify the algebra by confining the geometry to the plane of motion, where the coordinates of the point in motion are ξ and η (see [1]). Eq. (1.1) is then equivalent to the system

$$\ddot{\xi} + K\frac{\cos\varphi}{r^2} = 0, \quad \ddot{\eta} + K\frac{\sin\varphi}{r^2} = 0 \tag{1.2}$$

with r and φ the polar coordinates of the plane with respect to the attraction center. The magnitude of the rate of area swept by the position vector of the particle is then given by

$$\dot{a} \equiv \xi\dot{\eta} - \eta\dot{\xi} = r^2\dot{\varphi}. \tag{1.3}$$

This constant of motion allows us an elegant integration of the system (1.2) with the analytical form of the trajectory as a direct outcome. First we define the complex variable

$$z \equiv \xi + i\eta = re^{i\varphi}. \tag{1.4}$$

so that (1.2) can be written in the form

$$\ddot{z} + K\frac{e^{i\varphi}}{r^2} = 0 \tag{1.5}$$

Now, use (1.3) to eliminate, such that

$$\ddot{z} + K\frac{e^{i\varphi}}{\dot{a}}\dot{\varphi} = 0, \dot{z} = i\left(K\frac{e^{i\varphi}}{\dot{a}} + w\right), \tag{1.6}$$

where $w \equiv w_1 + iw_2$ is a complex constant of integration to be determined by *the initial conditions of the problem.* The analytical equation of motion can be then extracted directly from (1.3) by using (1.6). In polar coordinates of the plane of motion the result is

$$\frac{\dot{a}}{r} = \frac{K}{\dot{a}} + w_1 \cos\varphi + w_2 \sin\varphi. \tag{1.7}$$

The shape of this trajectory is best pictured by going back to Cartesian coordinates, where we have, instead of (1.7) the second-degree curve a conic:

$$\left(\frac{K^2}{\dot{a}^2} - w_1^2\right)\xi^2 - 2w_1w_2\xi\eta + \left(\frac{K^2}{\dot{a}^2} - w_2^2\right)\eta^2 + 2\dot{a}(w_1\xi + w_2\eta) = \dot{a}^2. \tag{1.8}$$

The center of this conic *is not the center of the force*, but has the coordinates

$$\xi_0 = -\frac{\dot{a}w_1}{\Delta}, \eta_0 = -\frac{\dot{a}w_2}{\Delta}, \Delta \equiv \left(\frac{K}{\dot{a}}\right)^2 - w_1^2 - w_2^2. \tag{1.9}$$

In cases where $\Delta = 0$, the center of this trajectory is at infinity: the trajectory is a parabola. We have here the ballistic cases, where the basic motion is parabolic.

Assuming the center of the trajectory at finite distance with respect to the center of force, and referring the trajectory to this center by the translation $x = \xi - \xi_0$, $y = \eta - \eta_0$, its equation becomes

$$\left(\frac{K^2}{\dot{a}^2} - w_1^2\right)x^2 - 2w_1w_2xy + \left(\frac{K^2}{\dot{a}^2} - w_2^2\right)y^2 = \frac{K^2}{\Delta}. \tag{1.10}$$

The quadratic form the left hand side of this equation is completely characterized by the 2×2 matrix

$$\boldsymbol{A} \equiv \begin{pmatrix} \dfrac{K^2}{\dot{a}^2} - w_1^2 & -w_1 w_2 \\ -w_1 w_2 & \dfrac{K^2}{\dot{a}^2} - w_2^2 \end{pmatrix}. \tag{1.11}$$

The eigenvalues of this matrix are Δ and $K^2/\dot{a}^2$, with the corresponding eigenvectors

$$|e_1> = \begin{pmatrix} \cos\omega \\ \sin\omega \end{pmatrix}, |e_2> = \begin{pmatrix} -\sin\omega \\ \cos\omega \end{pmatrix}, w_1 \equiv |w|\cos\omega, w_2 \equiv |w|\sin\omega. \tag{1.12}$$

Thus, the orientation of trajectory in its plane is completely defined by the initial conditions of the motion. The magnitude "w" is proportional with the eccentricity "e" of trajectory. Indeed the semiaxes "a" and "b" are

$$a^2 = \frac{K^2}{\Delta^2}, b^2 = \frac{\dot{a}^2}{\Delta}, e^2 \equiv \frac{a^2 - b^2}{a^2} = (\frac{\dot{a}}{K}\boldsymbol{w})^2. \tag{1.13}$$

Thus, the initial conditions can actually be expressed only in terms of "contemporary" magnitudes allowing us to *forget about the past*:

$$w_1 = \frac{K}{\dot{a}} e \cdot \cos\omega, \quad w_2 = \frac{K}{\dot{a}} e \cdot \sin\omega. \tag{1.14}$$

This is essentially the observation that imposed the Newtonian explanation for the real planetary motions in terms of contemporary quantities. Indeed, a force is always contemporary, and the initial conditions of the motion, whatever they might be, are then to be read in some contemporary parameters of motion: area constant and eccentricity.

As it can be seen directly from Eqs. (1.13) one of the semiaxes can be imaginary, for $\Delta < 0$, in which case we have to do with hyperbolic trajectories. It is only in cases where $\Delta > 0$, that we have to do with elliptic trajectories, properly representing planetary motion. Along this line of reasoning, the parabolic trajectories are all characterized by points on the circle $\Delta = 0$, i.e.

$$\mu^2 + v^2 = 1, \mu \equiv e \cdot \cos\omega, v \equiv e \cdot \sin\omega, \tag{1.15}$$

and the *whole interior of this circle corresponds to all possible finite motions that a material point can have* around a center of force acting with a force inversely proportional to square of distance. This would mean that a planet would have infinitely many possible initial conditions we have to choose from. The fact is that the actual motion of a planet is perceived as if it had *unique initial conditions.* Any departure from this perception has always induced arguments about some actual perturbations acting on the planet. To a certain extent this is true: the discovery of Neptune is an example. However, as the history shows, it has not the touch of universality needed for the continuity of knowledge. Thus, we have to turn to the origin of the problem, and lead the reasoning along the following lines: Kepler motion has reality only as a snapshot, this is undeniable; it could not have been discovered otherwise. However the planetary motion is a succession of such snapshots, which have to be put together in order to make the whole thing. First of all, we have to find the time scale of such a snapshot, and that is hard. But we have another possibility, opened by the remarks just made above: there is an *a priori* metric geometry of the defining parameters of the snapshot, which are the *initial conditions* of the dynamical problem describing this snapshot. This geometry defines a kinematics, and the kinematics offers us a natural way to continuously connect the snapshots in a succession representing a real trajectory.

Indeed, even superficially it can be seen at once that the mentioned freedom of the parameters defining the types of orbits, allows us to construct a Cayley-Klein (or Absolute) geometry [2, 3] characterizing the variation of those orbits. We know that an Absolute geometry is related to some conservation laws, at least as long as some realizations of $SL(2,R)$ group structure are involved. And indeed, the absolute metric for the interior of the circle (15)

$$(ds)^2 = \frac{(1-v^2)(d\mu)^2 + 2\mu v(d\mu)(dv) + (1-\mu^2)(dv)^2}{(1-\mu^2-v^2)^2} \tag{1.16}$$

can be brought to the form of Poincaré metric

$$(ds)^2 = -4\frac{dh \cdot dh^*}{(h-h^*)^2} = \frac{(du)^2+(dv)^2}{v^2} \tag{1.17}$$

by the following transformation of coordinates:

$$\mu = \frac{hh^*-1}{hh^*+1}, v = \frac{h+h^*}{hh^*+1} \Leftrightarrow h \equiv u+iv = \frac{v+i\sqrt{1-\mu^2-v^2}}{1-\mu}. \tag{1.18}$$

The conservation laws for the metric (1.17) are

$$\omega_1 = \frac{du}{v^2}, \omega_2 = 2\frac{udu + vdv}{v^2}, \omega_3 = \frac{(u^2 - v^2)du + 2uvdv}{v^2}. \tag{1.19}$$

The description we are also interested in now is the one in variables (e, ω), i.e. the eccentricity and the orientation of the orbit in its plane. In terms of these parameters the metric (1.16) becomes

$$(ds)^2 = \left(\frac{de}{1 - e^2}\right)^2 + \frac{e^2}{1 - e^2}(d\omega)^2. \tag{1.20}$$

We can rewrite this metric in a well-known form, by noticing that for elliptic trajectories "e" is confined to the interval between 1 and +1, so that the change of parameter

$$e = \tanh \Psi \tag{1.21}$$

is legitimate. With this the metric (1.20) becomes

$$(ds)^2 = (d\Psi)^2 + \sinh^2 \Psi (d\omega)^2. \tag{1.22}$$

The complex parameter "h" from equation (1.18) has a direct relationship with the theory of classical potentials. In order to show this relationship we write here "h" in terms of (e, ω). We have:

$$h = i\frac{\cosh \chi + \sinh \chi e^{-i\omega}}{\cosh \chi - \sinh \chi e^{-i\omega}}, \quad \aleph \equiv \frac{\Psi}{2}. \tag{1.23}$$

It just happens that this equation represents a harmonic map from the usual space into the Lobachevsky plane provided χ (and therefore ψ) is a solution of the Laplace equation in free space.

Indeed, the problem of harmonic correspondences between space and the hyperbolic plane is described by the minimum of energy functional corresponding to the metric (1.17) where the differentials are transformed into space gradients [4, 5]. The minimization of energy functional corresponds to Euler-Lagrange equations for the Lagrangian

$$\Lambda \equiv -4\frac{\nabla h \cdot \nabla h^*}{(h - h^*)^2}. \tag{1.24}$$

These are

$$(h - h^*) \cdot \nabla^2 h - 2(\nabla h)^2 = 0 \tag{1.25}$$

and its complex conjugate. Then it is easy to see that "h" from (1.23) verifies this equation when χ is a solution of Laplace equation, and ω does not depend on the position in space. It might, nevertheless, depend on the local time of the Newtonian dynamics.

This method can be thought of really ascribing "a spatial expanse", in the form of harmonic surfaces in space, of the regions of space extended over the ranges of eccentricity of the Kepler motion. This is to say, that the harmonic maps from the Lobachevsky plane to space are related to the physics of Sun, in the case of planetary system, or to the physics of nucleus in the case of classical atomic model. There is no obvious sign today in physics for the first case, i.e. the solar system, but the conclusion seems to be fair for the case of atomic model.

1.2 Some variations on a theme of skyrmions

We don't need to insist again in recalling that the main scientific image of the atom is the planetary one, amended perhaps with the idea that this physical system is not plane but a spatial one maybe spherical. This sphericity of the model, if real, is due to the noncentrality of forces or to the space extension of the matter at the nuclear level. If this is the case, then the region of the nucleus can be characterized by a 3D hyperbolic space. With this statement we enter the realm of recent date of the hyperbolic skyrmions [6, 7].

Indeed, one of the dominant contemporary concepts in the theory of structure of the nuclear matter, is that of skyrmion, which represents itself a variation on the subject of harmonic applications. The Skyrmion is a soliton representing the nucleons. The history of this subject starts with the physicist Tony Skyrme, and the reader in need of following it can begin with his recollections of the beginning [8], where the reasons and the original works are indicated. The main mathematical point of Skyrme's idea, is a certain, "almost harmonic" map, from the usual space to sphere, built after the traditional manner of the variational problem leading to the Laplace and Schrdinger equations. Let's shortly describe this manner, following by and large the recent work of Slobodeanu [9].

As known, such a harmonic map is obtained by finding the functions which realize the extremum of the energy functional

$$E_2(\mathbf{\Phi}) = \iiint \langle d\mathbf{\Phi} | d\mathbf{\Phi} \rangle d^3\mathbf{x}. \tag{1.26}$$

Here, the function realizing the correspondence is generally of the nature of a matrix; this is why we even denoted it by $\mathbf{\Phi}$ and the dot product is the one induced by the metric of the hyperbolic space. In the case of Lobachevsky plane, the metric Lagrangian (1.24) corresponds to the metric of the plane:

$$\mathbf{\Phi} \equiv \begin{pmatrix} h \\ h^* \end{pmatrix}; \quad \langle d\mathbf{\Phi}|d\mathbf{\Phi}\rangle \equiv -4\frac{dhdh^*}{(h-h^*)^2}. \tag{1.27}$$

This variational principle is directly connected with the inertial field in vacuum, because it is equivalent with Einstein vacuum field equations [10-12]. Consequently, those very equations can be simply considered as equally describing the states of nuclear matter, provided this one admits a description in the hyperbolic plane. And this is what the classical Kepler problem actually points out. One would therefore expect that based on that classical problem of motion, the solution of problem of the nuclear structure should be simply a matter of study of the harmonic maps, but this is not quite the case. The nucleus is assumed to have a particulate structure, and the harmonic principle, if not somehow amended, cannot account for that. So, it comes that the Skyrme functional is not quite as simple as the above one, but involves higher degree terms, belonging to different cohomology classes, given by the equation

$$E_4(\mathbf{\Phi}) = \iiint \langle d\mathbf{\Phi} \wedge d\mathbf{\Phi}|d\mathbf{\Phi} \wedge d\mathbf{\Phi}\rangle d^3\mathbf{x}. \tag{1.28}$$

If $\mathbf{\Phi}$ is a mapping from the Euclidean space to itself, then this term assures that the resulting equations of motion are nonlinear, and the nonlinear equations always admit confined (soliton) solutions, as the structure of the nucleus seems to require.

In order to establish the place of the hyperbolic geometry in the theory of nuclear matter, and therefore of the stellar matter or of some other nature, we just need to guide our guesses. The first observation is actually the gist of the work from 2004 of Atiyah and Sutcliffe [7], and is referring to the very significance of hyperbolic skyrmions. Namely, the theory of Euclidean skyrmions, with massive pions, leads to detailed results which are almost identical with those referring to massless hyperbolic skyrmions. It is as if the mass the source of inertia in classical mechanics and general relativity is somehow related to the curvature again an idea of general relativity not in the space-time but simply in space.

However, it seems more comfortable to work with maps from real space to itself like always in classical mechanics as in the cases leading to the equations of Laplace or Schrödinger. Along this line an essential observation has been made by N. S. Manton [13], and presents the Euclidean skyrmions as related to the deformations of matter. This is an idea of classical inspiration, taking its roots from the deformation theory, where the experimental deformations are described by the so-called tensor of elongations [14, 15]. The most general energetic functional of Skyrme can be written, according to Manton, in the form given in equation (1.26), where $\mathbf{\Phi}$ is now a vector dictated by the deformation of matter in the regular space:

$$\mathbf{\Phi} = \begin{pmatrix} \lambda_1 & 0 & 0 \\ 0 & \lambda_2 & 0 \\ 0 & 0 & \lambda_2 \end{pmatrix} \begin{pmatrix} x \\ y \\ z \end{pmatrix} \tag{1.29}$$

with $\lambda_{1,2,3}$ the elongations along three reciprocally orthogonal directions. Here the interior product from (1.26) is simply the dot product of regular vectors. As from equation (1.29) we have

$$d\mathbf{\Phi} = \begin{pmatrix} \lambda_1 & 0 & 0 \\ 0 & \lambda_2 & 0 \\ 0 & 0 & \lambda_2 \end{pmatrix} \begin{pmatrix} dx \\ dy \\ dz \end{pmatrix}, \tag{1.30}$$

it is plain that the energy functional from equation (1.26) is

$$E_2(\mathbf{\Phi}) = \iiint (\lambda_1^2 + \lambda_2^2 + \lambda_3^2) d^3\mathbf{x}. \tag{1.31}$$

As to the other term from the original theory of Skyrme, it is simply the dot product of the exterior square of the gradient:

$$d\mathbf{\Phi} \wedge d\mathbf{\Phi} = \lambda_2\lambda_3 dy \wedge dz + \lambda_3\lambda_1 dz \wedge dx + \lambda_1\lambda_2 dx \wedge dy. \tag{1.32}$$

So the functional from equation (1.28) can be written in the form

$$E_4(\mathbf{\Phi}) = \iiint (\lambda_2^2\lambda_3^2 + \lambda_3^2\lambda_1^2 + \lambda_1^2\lambda_2^2) d^3\mathbf{x}. \tag{1.33}$$

Incidentally, the indices 2 and 4 occuring in the energetic functionals of Manton are justified by the orders of the tensors entering the integrands of those functionals. Thus, $d\mathbf{\Phi}$ is defined by a second order tensor, while $d\mathbf{\Phi} \wedge d\mathbf{\Phi}$ is defined by a fourth order tensor.

Finally the Manton functional for the Skyrme model is the sum of the two contributions to the energy, i.e.:

$$E_4(\Phi) = \iiint (\lambda_1^2 + \lambda_2^2 + \lambda_3^2 + \lambda_2^2\lambda_3^2 + \lambda_3^2\lambda_1^2 + \lambda_1^2\lambda_2^2)d^3\mathbf{x}. \tag{1.34}$$

Details on this line of reasoning can be found in the original work of N. S. Manton [13], and in numerous works that followed along this line (see [16]). The interested reader is directed to those works, for our line of reasoning will take here another turn in the pursuit of hyperbolic skyrmions.

1.3 A geometry of hyperbolic skyrmions

The way we see it, the theory of hyperbolic skyrmions is the legitimate descendant of the classical theory describing the atomic structure based on the classical dynamics of electric forces. Indeed, according to our presentation above, this theory describes the region of atomic space normally assigned to the nucleus of the atom, and the geometry of this region is the hyperbolic geometry. In spite of the fact that Manton's idea is about Euclidean skyrmions, it contains almost explicitly a connection with hyperbolic skyrmions. This connection comes about in another work of Atiyah and Manton [17], whereby the deformation is treated in terms of the roots of a family of cubic equations.

The idea of a family of cubic equations involved in the geometry of skyrmions is indeed germane to the problem, as Atiyah and Manton show. However, it is also germane to the problem of deformation, inasmuch as a state of deformation is described in terms of matrices, and a cubic equation is simply the characteristic equation of a 3×3 matrix. A process of deformation is then described by a family of matrices, therefore by a family of cubic equations. And a family of cubic equation is always described by a metric of constant negative curvature that generalizes the metric of hyperbolic plane [18]. Thus we can come directly to a metric describing the deformation, and then use the harmonic principle in order to describe the hyperbolic skyrmions. Let us expound along this line.

From energetical point of view the Manton's functional is actually a very special one. Indeed, in the realm of hyperelastic deformations, it is only a special instance of the so-called Mooney-Rivlin model, where the energy density is a linear combination of the two invariants of the deformation

$$I_1 = \lambda_1^2 + \lambda_2^2 + \lambda_3^2; \quad I_2 = \lambda_2^2\lambda_3^2 + \lambda_3^2\lambda_1^2 + \lambda_1^2\lambda_2^2. \tag{1.35}$$

This model describes correctly deformations up to 30 - 40% which, for some rubbers, are even small deformations. However, it is the only theoretical model accepting a microscopic physical description by means of a Gaussian statistics of the macromolecular chains, so that the contemporary theoretical physics almost got stuck within its limits. Another reason for this might also be the fact that the range of deformations covered by this description is quite enough for application in phenomena with industrial application. However, theoretically speaking, the physics cannot afford to say that this model universally describes the process of deformation: it cannot be taken as a fundamental law of nature. So much less, therefore, can we say that such a model will describe the deformations of nuclear matter. Here, we are in the hazy range of the confluence between the matter proper and space. It is to be expected, for instance, that the deformations of nuclear matter are not reversible, because of the energy dissipation in the form of particles and heat, like, for instance inside the Sun our daily nucleus.

The general idea here would be that the irreversible deformations taking place with dissipation of energy, are dominated not by the usual invariants of deformation, but by an algebraical combination of them, appearing as an intensity of shearing deformations, as Novozhilov [19] has shown. This algebraical combination carries, in the theory of deformations, the name of invariant of von Mises. In terms of the invariants from equation (1.35) it can be written as

$$\frac{1}{2}(I_1^2 - 3I_2) \equiv (\lambda_2^2 - \lambda_3^2)^2 + (\lambda_3^2 - \lambda_1^2)^2 + (\lambda_1^2 - \lambda_2^2)^2. \tag{1.36}$$

Consequently, the Manton-like functional corresponding to this situation should be written in the algebraically homogeneous form

$$E(\mathbf{\Phi}) = \frac{1}{2}\iiint \{\langle d\mathbf{\Phi}|\rangle^2 - 3\langle d\mathbf{\Phi} \wedge d\mathbf{\Phi}|d\mathbf{\Phi} \wedge d\mathbf{\Phi}\rangle\} d^3\mathbf{x}, \tag{1.37}$$

which is, obviously, out of the limits of Skyrme theory as first conceived. However, this functional suggests a consistent way of description of the nuclear matter, within the limits of general relativity, inasmuch as it generalizes the Ernst approach. This consists of reducing the Skyrme functional to its first term only, therefore to the usual energy of a harmonic application. We will work therefore on the form of the map $\mathbf{\Phi}$ itself, and generalize it in a natural manner, so as to include the regular hyperbolic geometry as a special case. In order to understand this extension a short incursion in a special theory of deformation of higher orders will be necessary, suggested by the Barbilians cubic space [18] considered as a Cayley-Klein space.

Let us therefore make only the assumption that the deformations are represented by a 3×3 matrix, without being interested if it is indeed derivable from a gradient or not. It can represent, for instance, a variation, at a certain scale, of the metric tensor of the space within which the matter is contained. Let us denote by "x" this deformation matrix. Its eigenvalues, ($(\lambda^2 - 1)/2$ in the previous formalism of Manton) are the roots of a third degree equation, the characteristic equation of matrix. This is the circumstance allowing us to write a general deformation energy density, as based on the algebraical principle of the polarity of the binary algebraic forms. This will be explained as we proceed.

Assume, to start with, that we have a quadratic binary form - a homogeneous second degree polynomial - having the coefficients a_0, a_1 and a_2, which happens to have some physical meaning in a physical problem. Assume also that we have a set of cubic binary forms, representing the characteristic polynomials of our deformation matrices. These cubic polynomials have a common invariant with our starting quadratic polynomial [20]. This invariant is a quadratic form in the coefficients of our family of cubic polynomials:

$$\frac{1}{9}\Psi \equiv a_2(X_0X_2 - X_1^2) - a_1(X_0X_3 - X_1X_2) + a_0(X_1X_3 - X_2^2). \quad (1.38)$$

Here, X_0, X_1, X_2 and X_3 are the coefficients of a generic cubic from our family, taken in its binomial form

$$X_0x^3 + 3X_1x^2 + 3X_2x + X_3. \quad (1.39)$$

X_0 the coefficient of the third degree term of the polynomial, is responsible for the indetermination in the relations between roots end coefficients, and x is the generic nonhomogeneous variable of the cubic. Vanishing of the quadratic form (1.38) implies the apolarity between our starting quadratic form and every member of the family of cubics. As known, the apolarity can be extended to a projective concept, which here comes in handy, inasmuch as the ratios

$$\frac{X_1}{X_0}, \frac{X_2}{X_0}, \frac{X_3}{X_0} \tag{1.40}$$

can be taken as nonhomogeneous coordinates of a point in a 3D space of cubics.

From the point of view of the deformation process, let's first notice that the function Ψ from equation (1.38) can be taken as a potential which generalizes, in a natural manner, the function from equation (1.36) to a nonhomogeneous function. Indeed, that function is given by the first term only, from the right hand side of equation (1.38). The theory of potential works here directly, as follows. First, if the generic cubic of our family is the characteristic equation of the symmetric matrix , with elements representing deformations, then the matrix having the entries given by equation

$$y_{ij} = \frac{\partial \Psi}{\partial x_{ij}}, \tag{1.41}$$

represents the corresponding stresses; obviously, the opposite is also valid. Equation (1.41) suggests that Ψ is a potential indeed, according to the rules of using the concept of potential.

In our conditions the equations (1.41) should be necessarily of the form

$$y_{ij} = \alpha_0 \delta_{ij} + \alpha_1 x_{ij} + \alpha_2 (\mathbf{x}^2)_{ij}, \tag{1.42}$$

because they correlate 3×3 matrices. The problem is to find the coefficients , which represent the physical properties of the continuum whose deformation is described by the matrix x. In order to solve this problem we need to have the coordinates X_j which, up to a common factor are given by the coefficients of the characteristic equation of x:

$$X_0 = 1, \;\; X_1 = -\frac{I_1}{3}, \;\; X_2 = \frac{I_2}{3}, \;\; X_3 = -I_3. \tag{1.43}$$

Here, I_1, I_2 and I_3 denote the scalar quantities specific to the matrix x, that in the case of a Euclidean tensor are its orthogonal invariants. In general though, they are defined by the formulas

$$I_1 = Tr(x) \equiv t_1, \;\; I_2 = \frac{1}{2}[(Tr(x))^2 - Tr(x^2)] \equiv \frac{1}{2}(t_1^2 - t_2),$$
$$I_3 = \frac{1}{6}[2Tr(x^3) - 3(Tr(x))Tr(x^2) + (Tr(x))^3] \equiv \frac{1}{6}(2t_3 - 3t_1 t_2 + t_1^3), \tag{1.44}$$

where we used obvious notations for the traces of powers of the matrix. Using these we have

$$\begin{aligned} A_0 &\equiv X_0X_2 - X_1^2 = \frac{1}{18}(t_1^2 - 3t_2), \\ A_1 &\equiv X_0X_3 - X_1X_2 = -\frac{1}{9}(t_1^3 - 4ty_1t_2 + 3t_3), \\ A_2 &\equiv X_1X_3 - X_2^2 = \frac{1}{36}(4t_1t_3 - 4t_1^2t_2 + t_1^4 - t_2^2). \end{aligned} \tag{1.45}$$

Therefore, in order to calculate (1.41) we need the derivatives of the traces of different powers of the matrix x. These are calculated according to the formulas

$$\frac{\partial t_1}{\partial x_{ij}} = \delta_{ij}, \quad \frac{\partial t_2}{\partial x_{ij}} = 2x_{ij}, \quad \frac{\partial t_3}{\partial x_{ij}} = 3(x^2)_{ij}, \tag{1.46}$$

so that, finally, we have for α_0, α_1 and α_2 the following equations:

$$\begin{aligned} \alpha_0 &= a_2t_1 + a_1(3t_1^2 - 4t_2) + a_0(t_1^3 - 2t_1t_2 + t_3), \\ \alpha_1 &= -3a_2 - 8a_1t_1 - a_0(2t_1^2 + t_2), \\ \alpha_2 &= 3(3a_1 + a_0t_1). \end{aligned} \tag{1.47}$$

On the other hand the constitutive relation (1.42) is also inversible, i.e. we can write the deformations in terms of stresses in the form

$$\mathbf{x} = \beta_0\mathbf{e} + \beta_1\mathbf{y} + \beta_2\mathbf{y}^2, \tag{1.48}$$

where e is the identity 3×3 matrix. β_1, β_2 and β_3 are here connected with α_0, α_1 and α_2 by the system of equations

$$\begin{aligned} &\beta_0 + \beta_1\alpha_0 + \beta_2(\alpha_0^2 + \alpha_2^2I_1I_3 + 2\alpha_1\alpha_2I_3) = 0, \\ &\alpha_1\beta_1 + [2\alpha_0\alpha_1 - 2\alpha_1\alpha_2I_2 + (I_1^2 - I_2)\alpha_2^2] \cdot \beta_2 = 1, \\ &\alpha_2\beta_1 + [\alpha_1^2 + 2\alpha_1\alpha_2I_1 + 2\alpha_0\alpha_2 + (I_3 - I_1I_2)\alpha_2^2] \cdot \beta_2 = 0. \end{aligned} \tag{1.49}$$

This shows that there are nontrivial states of stress, corresponding to null deformations, as in the case of vacuum for instance, or that of classical ether. These are determined exclusively by the starting binary quadratic form, through the equations

$$\begin{aligned} &\beta_0 + \beta_1\alpha_0 + \beta_2\alpha_0^2 = 0, \\ &\alpha_1\beta_1 + 2\alpha_0\alpha_1\beta_2 = 1, \\ &\alpha_2\beta_1 + (\alpha_1^2 + 2\alpha_0\alpha_2)\beta_2 = 0. \end{aligned} \tag{1.50}$$

where, according to (1.47), we take

$$\alpha_0 = 0, \quad \alpha_1 = -a_2, \quad \alpha_2 = 9a_1. \tag{1.51}$$

Thus, we have

$$\beta_0 = 0, \quad \beta_1 = -\frac{1}{3a_2}, \quad \beta_2 = \frac{a_1}{3a_2^3}. \tag{1.52}$$

Our starting quadratic form should therefore represent such limit states of matter, that do not depend on the state of deformations or tensions. Generally, however, the constitutive relations (1.42) and (1.48) cannot be simultaneously inversed, because there are separately deformations that do not involve stresses, as well as stress phenomena that do not involve deformations. The matter for which these phenomena are simultaneous should therefore have special properties.

An example of such states can be given immediately. Indeed, the relation (1.42) shows that the stresses can all vanish for a certain nontrivial deformation processes. Indeed, the system of equations

$$\alpha_0 = \alpha_1 = \alpha_2 = 0, \tag{1.53}$$

has a nontrivial solution for the invariants of deformations. This solution is given by the following formulas for the traces of powers of deformation matrix as functions of the coefficients of the starting binary quadratic

$$t_1 = -3\frac{a_1}{a_0}, \quad t_2 = 3\frac{2a_1^2 - a_0a_2}{a_0^2}, \quad t_3 = 3\frac{4a_1^3 - 3a_0a_1a_2}{a_0^3}. \tag{1.54}$$

Therefore, the eigenvalues of deformation matrix are functions independent of the state of stress. In cases where the starting quadratic form has real roots, say $u \pm v$, these equations give

$$t_1 = 3u, \quad t_2 = 3(u^2 + v^3), \quad t_3 = 3u(u^2 + 3v^2). \tag{1.55}$$

Using now the formulas (1.43) and (1.44) we find easily the coefficients of the characteristic equation of the corresponding matrix in the form

$$\frac{c_0}{1} = \frac{c_1}{-u} = \frac{c_2}{u^2 - \frac{v^2}{2}} = \frac{c_3}{-u(u^2 - \frac{3v^2}{2})}. \tag{1.56}$$

This cubic is the center of the quadric given in equation (1.38), and has the remarkable property of having the roots strictly determined by u and v:

$$u - \sqrt{\frac{3}{2}}v, \quad u, \quad u + \sqrt{\frac{3}{2}}v. \tag{1.57}$$

The Hessian of this reference cubic is given by the equations

$$\frac{A_0}{1} = \frac{A_1}{-2u} = \frac{A_2}{u^2 + \frac{v^2}{2}} = -\frac{v^2}{2}. \tag{1.58}$$

Obviously, this is a polar conjugate of our starting binary quatratic. There are also a family of cubics conjugated with (1.58) in the sense that their Hessians are apolar with this one. This is the original Barbilian hyperboloid, having the equation:

$$(u^2+\frac{v^2}{2})(X_0X_2-X_1^2)+u(X_0X_3-X_1X_2)+(X_1X_3-X_2^2) = 0. \tag{1.59}$$

The center of this quadric is given by

$$\frac{b_0}{1} = \frac{b_1}{-u} = \frac{b_2}{u^2 + \frac{v^2}{2}} = \frac{b_3}{-u(u^2 + \frac{3v^2}{2})} \tag{1.60}$$

and obviously is a cubic having only one real root, i.e. its roots are

$$u - i\sqrt{\frac{3}{2}}v, \quad u, \quad u + i\sqrt{\frac{3}{2}}v. \tag{1.61}$$

Therefore we have here a kind of duality. One can say that and are functions of the physical characteristics of matter undergoing deformation, and that there are states of strain and stress that do not depend on anything else but on these physical characteristics. This is the meaning of the starting quadratic form, which represents the Hessian of a certain cubic form.

The quadric from equation (1.59) is the starting point of Dan Barbilian in the construction of the Riemann spaces associated with families of one-parameter cubics, as Cayley-Klein spaces [18]. The geometrical procedure used by Barbilian will be now discussed in broad strokes. First, we need to notice that Barbilian begins with the idea that the starting quadratic form is the Hessian of a cubic with real roots, therefore it has complex roots itself. Then, by performing the linear transformation of homogeneous coordinates

$$\frac{X_0'}{uX_0 + X_1} = \frac{X_1'}{u^2X_0 - X_2} = \frac{X_2'}{(u^2 + \frac{v^2}{2})X_0 - X_2}$$
$$= \frac{X_3'}{u(u^2 + \frac{v^2}{2})X_0 - (u^2 + \frac{v^2}{2})X_1 - uX_2 + X_3} \quad (1.62)$$

we can reduce (1.59) to a "canonical" form

$$X_0'X_3' - X_1'X_2' = 0. \quad (1.63)$$

showing explicitly that we are dealing here with a one-sheeted hyperboloid. Using now the Sylvester theorem, for the representation of a cubic in terms of its Hessian [20], we get the interesting result that a cubic from the family having the same Hessian can be represented by a 2×2 matrix having the entries

$$\frac{X_0'}{h^*k} = \frac{X_1'}{h} = \frac{X_2'}{k} = \frac{X_3'}{1}. \quad (1.64)$$

where h and h^* are the complex roots of the Hessian, and k is an arbitrary complex factor of unit modulus. The Cayley-Klein metric of this representation, with respect to the hyperboloid from equation (1.63) as an Absolute, is the Barbilian metric given by

$$(ds)^2 = -4\frac{dhdh^*}{(h-h^*)^2} + \left(\frac{dk}{k} - \frac{dh + dh^*}{h - h^*}\right)^2. \quad (1.65)$$

The phase factor can be interpreted in terms of the ensembles of harmonic oscillators [21]. Geometrically it is the parameter of a family of Bäcklund transformations from the Lobachevsky geometry, characterized by the last part of the metric (1.65), to a general 3D hyperbolic geometry characterized by the whole metric (1.65). It is thus to be expected that the harmonic mapping associated with the energy functional

$$E(\Phi) = \frac{1}{2}\iiint\left\{-4\frac{\nabla h \nabla h^*}{(h-h^*)^2} + \left(\frac{\nabla k}{k} - \frac{\nabla h + \nabla h^*}{h - h^*}\right)^2\right\} d^3\mathbf{x} \quad (1.66)$$

will give us skyrmions, thus bridging the gap between general relativity and the theory of nuclear matter. We even have the possibility of an anzatz here.

Indeed, we have seen that the first part of the metric (1.65) is the one generating the energetic functional for the gravitational field in vacuum, and the variational principle associated to it, leads to Ernst's equations, equivalent to Einstein's equations in vacuum. If we express the complex number h as a function of the eccentricity e of the orbit, by the formula (1.21), this metric will be given by equation (1.22). This is the metric of a section of hyperbolic space, used by Atiyah and Sutcliffe in the construction of hyperbolic skyrmions. Its form is, in general

$$(ds)^2 = (d\psi)^2 + \sinh^2 \psi (d\theta^2 + \sin^2 \theta d\varphi^2), \tag{1.67}$$

where θ and φ are usual spherical polar angles. Obviously, (1.22) can be obtained from (1.67) if we agree that ω represents the geodesic arc on the unit sphere. But (1.67) is also the absolute metric of the space of relativistic velocities [22], and can be directly obtained from (1.65), and probably there are still many meanings of it. The fact that it was used in representing the skyrmions with zero mass pions, motivates us to construct for it a skyrmion with distinguished significance in the Newtonian mechanics.

From the point of view of Newtonian forces, it is quite probable that the third principle of dynamics is no more effective over the space of nuclear matter, and that the eccentricity of the electronic orbits is actually an expression of the nonequilibrium of the forces. Here the Newtonian measurements of the forces one by means of another can still be defined, but by means of a quantum definition of the measurement, in the axiomatic manner in which the spin is introduced within quantum mechanics [23]. Namely, the 2×2 matrix

$$\mathbf{Q} \equiv \begin{pmatrix} \cos\theta & \sin\theta e^{i\varphi} \\ \sin\theta e^{-i\varphi} & -\cos\theta \end{pmatrix} \tag{1.68}$$

has eigenvalues ± 1, or any two numbers equal and opposite in sign, which is the ideal case of classical measurement of forces: equal and opposite. Consequently it can represent the third principle of dynamics (ad literam), in the sense that the two eigenvalues are the values of a force and its reaction, i.e. their algebraic sum is zero. Based on this, we can build a matrix that has two different eigenvalues, representing two different forces no matter of their directions. Indeed, any 2×2 matrix of the form

$$\mathbf{M} = \lambda \mathbf{E} + \mu \mathbf{Q}, \tag{1.69}$$

where λ and μ are real, and $\mathbf{E}$ is the 2×2 identity matrix, has two different real eigenvalues not depending on the angles. These are $(\lambda \pm \mu)$. Our ansatz is then identical with the original one of Skyrme, and amounts to an exponential expression of the matrix $\mathbf{M}$:

$$\mathbf{M} = \exp(\psi \mathbf{Q}). \tag{1.70}$$

One can really see that this means that the two forces are always in the same ratio, no matter of how their system is oriented in space. The absolute metric of the matrices (1.70) is just the metric from equation (1.67), where we are to take

$$\frac{\mu}{\lambda} = \tanh \psi. \tag{1.71}$$

Consequently the ratio of forces is practically represented by the eccentricity of the Keplerian motion which represents the atom. The "hedgehog" skyrmion from the equation (1.70) which we would like to call fundamental then represents the situation of forces at the nuclear level. They are in equilibrium according to the third law of Newton, only when the eccentricity of the electronic orbits is zero, i.e. when these orbits are perfect circles, as it should be according to the classical theory of Newtonian forces. This fundamental skyrmion is, according to the ideas of Atiyah and Sutcliffe, essential in the construction of any other with null mass pions.

1.4 Perspectives

There are reasons to hope that the theory of nuclear forces can be naturally unified with the general relativity (for details, see Refs. [24-27]. First, the existence of nucleus is quite naturally described even within the classical theory of Kepler motion, by the same harmonic mapping as that from general relativity. Then, this very description is quite close to a successful theory of the nuclear matter - the Skyrme theory (for alternative interpretations of this theory, see Refs. [28-32]. Only, the nuclear matter, and probably the matter in general, should be described not by an Euclidean geometry but by a Lobachevsky geometry. The way we see it now points toward the fact that the geodesics of this hyperbolic geometry represent motions of continuum matter giving out the energy, or energies, we usually notice. Then the affine parameter of those geodesics should be the temperature of the matter. It plays here the same role the time plays in the Newtonian kinematics. For other details see [33].

References

1. Mittag, L., Stephen, M.J.: American Journal of Physics, 60, 207, (1992).
2. Cayley, A.: Philosophical Transactions of the Royal Society of London, 149, 61, (1859), (Reprinted in The Collected Mathematical Works, Vol. II (Cambridge University Press, Cambridge, 561, 1889)).
3. Klein, F.: Annales de la Faculté des Sciences de Toulouse Série 1 Tome 11, G1-G62 (1897) (French translation of the original German memoir from Mathematische Annalen, Vol. IV, 573, (1871)).
4. Eells, J., Sampson, J.H.: American Journal of Mathematics, 86, 109, (1964).
5. Misner, C.W.: Physical Review D, 18, 4510, (1978).
6. Atiyah, M.F., Sutcliffe, P.: The Geometry of Point Particles, arXiv:hep-th/01051794v2, (2001).
7. Atiyah, M.F., Sutcliffe, P.: Skyrmions, Instantons, Mass and Curvature, arXiv:hep-th/0411052v1, (2004).
8. Skyrme, T.H.R.: International Journal of Modern Physics, A3, 2745, (1988).
9. Slobodeanu, R.: On the Geometrized Skyrme and Faddeev Models, arXiv:0809.4864v5, (2009).
10. Ernst, F.J.: Physical Review, 167, 1175, (1968).
11. Ernst, F.J.: Journal of Mathematical Physics, 12, 2395, (1971).
12. Israel, F.J. W., Wilson, G.A.: Journal of Mathematical Physics, 13, 865, (1972).
13. Manton, N.S.: Communications in Mathematical Physics, 111, 469, (1987).
14. Hill, H.: Journal of the Mechanics and Physics of Solids, 16, 229, (1968).
15. Ogden, R.W.: Proceedings of the Royal Society of London, A326, 565, (1972).
16. Manton, N.S., Sutcliffe, P.M.: Topological Solitons, Cambrige University Press, Cambridge, (2004).

17. Atiyah, M.F., Manton, N.S.: Communications in Mathematical Physics, 152, 391, (1993).
18. Barbilian, D.: Comptes Rendus de l'Académie Roumaine des Sciences, 2, 345, (1938).
19. Novozhilov, V.V.: Prikladnaya Matematika i Mekhanika, 16, 617, (1952), (in Russian).
20. Burnside, W.S., Panton, A.W.: The Theory of Equations, Dover Publications, New York, (1960).
21. Mazilu, N.: A Case Against the First Quantization, viXra.org: Quantum Physics/1009.0005, (2010).
22. Fock, V.A.: The Theory of Space, Time and Gravitation, Macmillan Company, New York, (1964).
23. Schwartz, H. M.: International Journal of Theoretical Physics, 16, 249, (1977).
24. Sutcliffe, P.: Modern Physics Letters B 29, 16 (2015).
25. Chakravarty, S., Hsu, C.-H.: Modern Physics Letters B 29, 16 (2015).
26. Ioannou, P.D., Diakonos, K.F., Maintas, X. N., Agop, M.: Modern Physics Letters A 28, 30 (2013).
27. Wang, J.B., Ren, J.R.: Modern Physics Letters B 23, 16 (2009).
28. Wu, C.: Modern Physics Letters B 23, 01 (2009).
29. Kim, K.S., Tanaka, A.: Modern Physics Letters B 29, 16 (2015).
30. Liu, Y.K., Zhang, C., Yang, S.J.: Modern Physics Letters B 27, 25 (2013).
31. Kaplunovsky, V., Melnikov, D., Sonnenschein, J.: Modern Physics, Letters B 29, 16 (2015).
32. Ma, Y.L., Harada, M., Lee, H.K., Oh, Y., Rho, M.: International Journal of Modern Physics: Conference Series, 29 (2014).
33. Mazilu, N., Agop, M., Gaţu, I., Iacob, D.D., Ghizdovăţ, D.: Modern Physics Letters B, 30, 13 (2016).

CHAPTER 2

Spherical metrics in general relativity

2.1 Principles of general relativity

The theory of gravitational fields, build up on the theory of relativity, is called general theory of relativity (GTR). This theory was created by Albert Einstein, on a purely deductive basis, and has later been proved by astronomical observations, as having a great amount of truth ([1]-[17]).

There is a main difference between gravitational fields, on the one hand, and the rest of the fields (electromagnetic, nuclear, etc.), on the other: under certain given initial conditions, all bodies - independently on their mass - perform one and the same motion. For example, the free fall laws in terrestrial gravitational field

$$\begin{aligned} v &= gt, \\ h &= \frac{g}{2}t^2, \\ v^2 &= 2gh \end{aligned} \tag{2.1}$$

are the same for all bodies, no matter what their mass is, and move with the same acceleration g.

This property allows one to establish a local equivalence between the inertial frames situated in gravitational field and those non-inertial. In other words, an inertial reference system is equivalent to a certain kind of gravitational field. For example, a uniformly accelerated frame is equivalent to a constant and homogeneous gravitational field [13]. On the contrary, a non-uniformly accelerated frame moving in a straight line is equivalent to a homogeneous, but variable gravitational field. All these facts lead to a postulate which in general theory of relativity is known as the equivalence principle.

Nevertheless, the non-inertial systems are not perfectly equivalent to the real gravitational fields: the essential difference appears in their behavior at infinity. The real gravitational field infinitely far from the producing sources always tends to zero, while in a rotating reference system the gravitational fields unlimitedly increase with moving off the rotation axis [13]. Let ds be the arc element of an inertial frame in cylindrical coordinates (r', φ', z'), given by

$$ds^2 = c^2 dt^2 - dr'^2 - r'^2 d\varphi'^2 - dz'^2. \tag{2.2}$$

If one passes to the rotating coordinate system (r, φ, z)

$$\begin{aligned} r' &= r, \\ \varphi' &= \varphi + \Omega t, \\ z' &= z, \end{aligned} \tag{2.3}$$

where Ω is the angular velocity, then (2.2) becomes

$$ds^2 = (c^2 - \Omega^2 r^2) dt^2 - 2\Omega r^2 d\varphi\, dt - dz^2 - r^2 d\varphi^2 - dr^2. \tag{2.4}$$

According to equation (2.4), the rotating frames can be used only up to distances

$$r < \frac{c}{\Omega}, \tag{2.5}$$

since for greater distances the rotation velocity of such reference frames would exceed the speed of light. This way, the postulate of general relativity concerning the existence of a speed limit c for propagation of any action, i.e. regardless of its nature, would become sensitive.

The mathematical formalism of the general relativity is Riemannian geometry. In constructing his theory, Einstein gave an analytical expression for the notion of distance, analogous to (but more general than) that furnished by the Pythagorean theorem, which is

$$ds^2 = g_{ik} dx^i dx^k; \quad i, k = 0, 1, 2, 3, \tag{2.6}$$

where g_{ik} are functions of spatial coordinates x_1, x_2, x_3 and temporal coordinate x_0.

The possibility of setting up a physical theory based on these facts is guaranteed by the following hypothesis: the trajectory of a material point in gravitational field defines the geodesic of Riemannian manifold with metric (2.6). This way, any gravitational perturbation shall act as a metric perturbation. In the general case of an arbitrary varying gravitational field, the metric is not only Riemannian but also time-varying, so that the relations between various geometric distances are also time-varying. As a consequence, the mutual space-disposal of the probe particles situated in gravitational field cannot remain unmodified, in any coordinate system. In other words, the general relativity cannot accept existence of mutually immobile systems of physical bodies.

The above exposed facts essentially modify the notion of reference frame in general relativity, as compared with the same notion met in classical mechanics, where by reference frame is understood a fixed material system with respect to which the motion is studied. Such material systems cannot exist in the presence of a variable gravitational field, so that to determine position in space of a particle one must have an infinite number of such systems to fill the space as a "medium". These material systems, together with arbitrarily working horologes connected to each of them, define in general relativity a reference frame [13]. This way, one appears necessity to introduce the principle of covariance of physical laws with respect to general coordinate transformations.

The real gravitational field cannot be eliminated by any coordinate transformation. Differently speaking, in the presence of gravitational field the space-time behaves in such a manner that by no transformation the quantities gik can be reduced to the Galilean form, that is

$$
\begin{aligned}
&g_{11} = g_{22} = g_{33} = 1, \\
&g_{00} = -1, \\
&g_{ik} = 0, \quad i \neq k.(1.7)
\end{aligned}
\tag{2.7}
$$

Nevertheless, there is a case when the gravitational field can be eliminated by a convenient coordinate transformation, namely the case of uniform gravitational field of a small spatial extension. In this case, gik reduces to (2.7), and such a coordinate system is a Galilean system for that given point.

2.2 Distance and synchronization in general theory of relativity

The metric (2.6) can also be written as [13]

$$ds^2 = g_{00}(dx^0)^2 + 2g_{0\alpha}dx^0dx^\alpha + g_{\alpha\beta}dx^\alpha dx^\beta, \quad \alpha, \beta = 1, 2, 3. \tag{2.8}$$

This expression introduces the three-dimensional tensor

$$\gamma_{\alpha\beta} = -g_{\alpha\beta} + \frac{g_{0\alpha}g_{0\beta}}{g_{00}}, \tag{2.9}$$

as well as the vector $\boldsymbol{g}$ of components

$$g_\alpha = -\frac{g_{0\alpha}}{g_{00}}. \tag{2.10}$$

Relation (2.8) shows the connection between the metric of real space and that of four-dimensional space-time and gives the squared distance element

$$dl^2 = \gamma_{\alpha\beta}dx^\alpha dx^\beta. \tag{2.11}$$

As we have seen, g_{ik} are time-dependent and, as a result, $\gamma_{\alpha\beta}$ are also time-dependent. Consequently, expression

$$\int dl \tag{2.12}$$

depends on the universe line and, consequently, makes no sense. Therefore, in general relativity, the notion of distance looses its general meaning, since it is only locally conserved. The only situation when the distance can still be defined is that with gik time-independent (constant gravitational field); in this case, expression (2.12) makes sense along any spacial curve.

Relation (2.9) interferes in calculation of difference of "time" x^0 for two simultaneous events that happen at two infinitely closed points. This takes the form

$$\Delta x^0 = g_\alpha dx^\alpha. \tag{2.13}$$

Under such a perspective, the horolodge synchronization reduces to the condition that 1-form (2.13) is a total differential of a function of spacial coordinates. In its tern, synchronization of horologes along a closed contour is - in general - impossible. Indeed, if the contour is covered only once, one finds for the expression

$$\oint g_\alpha dx^\alpha \tag{2.14}$$

a non-zero value. Mare than that, the univoc synchronization of the horologes in the whole space is impossible (except for the reference systems in which all the components of g_α are zero - synchronic frames) [13]. Impossibility of synchronizing all horologes is a property connected to the frame, and not to the space-time itself. Nevertheless, in a uniform gravitational field one can always choose a reference system in which the three quantities g_α be zero and, this way, make possible the complete synchronization of the horologes. The number of such procedures is infinite [13].

2.3 Constant gravitational field

A gravitational field is constant if one can choose a frame in which all components of the metric tensor g_{ik} are independent on the temporal coordinate x^0, called cosmic time of the metric.

In general, the choice of the cosmic time is not univoc. By adding to x_0 an arbitrary function of spatial coordinates, all components of g_{ik} shall still be expressed without x_0; this transformation corresponds to an arbitrary choice of the origin of cosmic time in each point of the space [13]. The cosmic time can also be multiplied by an arbitrary constant, without any change of the metric tensor g_{ik}; in this case, the choice if the time gauge is completely arbitrary.

Only the gravitational field produced by a single body can be a constant. As we have seen, in case of a system of many bodies their mutual attraction produces a variation in their fields [13].

If the body which generate the gravitational field is at rest in the frame with g_{ik} independent on x_0, then both time directions are equivalent. The time origin in all points of the space can be conveniently chosen, so as metric (2.6) remain unchanged when the sign of x_0 varies. As a result, all components $g_{0\alpha}$ of the metric tensor are null. We shall call these gravitational fields - static gravitational fields.

Immobility of a body is not a necessary condition for the constancy of its field. A body with axial symmetry rotating about its axis also produces a constant gravitational field. But in this last case the two time directions are not equivalent: the change of sign of the cosmic time in (2.4) leads to the change of the angular velocity [13]. Consequently, the components $g_{0\alpha}$ of the metric tensor corresponding to these fields are - in general - different from zero. We shall call these gravitational fields - stationary gravitational fields.

Therefore, if the rotating system is conceived as a horolodge, it indicates two moments of time for an event, and the one considered by nowadayss physics is connected with the notion of causality. This way could be given a correct solution to one of the most flaming problems in many-body dynamics: compatibility between motions of the bodies in the system.

The following investigation concerns analysis of several metrics associated with constant/stationary gravitational fields with spherical symmetry, by means of the above mentioned Einsteinian approach. Such an analysis of certain particular metrics puts into evidence some common points regarding formal treatment of the motion. Even if the Einsteinian point of view regarding the motion on geodesic sometimes proves to be limitative, the metric nevertheless remains a basic element. As it is shown by N. Ionescu-Pallas [9,10], even those models of universe defined by means of ingenious operational procedures - like the Milnes kinematic model [15] - can be reformulated by means of notion of metric.

2.4 The Schwarzschild metric. Non-null geodesics

In spherical coordinates r, θ, ϕ, the Schwaerzschild metric invariant writes

$$ds^2 = \left(1 - \frac{n}{r}\right) c^2 dt^2 - \left(1 - \frac{n}{r}\right)^{-1} dr^2 - r^2(d\theta^2 + \sin^2\theta d\varphi^2), \quad (2.15)$$

where r, t, n, c are, respectively, the "vector radius", the cosmic time, the gravitational radius, and the speed of light in vacuum. As gravitational radius one usually takes the value

$$n = \frac{2GM_0}{c^2}, \quad (2.16)$$

where G is the constant of gravitational interaction and M_0 the rest mass of the source-body.

If the motion of particles in gravitational field is performed on geodesics, then the equations of motion are obtained by means of the variational problem [8]

$$\delta \int ds = 0, \tag{2.17}$$

which is equivalent with

$$\delta \int L dt = 0, \tag{2.18}$$

where [13]

$$L dt + m_0 c ds = 0, \tag{2.19}$$

and L is the Lagrangian of the considered physical problem.

As we have mentioned, the invariant (2.15) respects the spherical geometry, and this fact is transfered mutatis mutandis to the Lagrangian. This means that our Lagrangian shall be invariant with respect to an orthogonal transformations of coordinates and velocities. This fact allows one to get close to the classical Kepler problems, while Binets equation leads to the plane solution. Invariance with respect to orthogonal transformations bring into discussion, indeed, the three areas first integrals [14].

Let x_1, x_2, x_3 be the Cartesian coordinates and $\dot{x}_1, \dot{x}_2, \dot{x}_3$ their derivatives with respect to time. The considered first integrals then are:

$$\begin{aligned} x_2 \frac{\partial L}{\partial \dot{x}_3} - x_3 \frac{\partial L}{\partial \dot{x}_2} &= C_1, \\ x_3 \frac{\partial L}{\partial \dot{x}_1} - x_1 \frac{\partial L}{\partial \dot{x}_2} &= C_2 \\ x_1 \frac{\partial L}{\partial \dot{x}_2} - x_2 \frac{\partial L}{\partial \dot{x}_1} &= C_3 \end{aligned} \tag{2.20}$$

where C_1, C_2, C_3 are constants of integration. After a convenient multiplication by x_1, x_2, x_3 in (2.20), we still have

$$C_1 x_1 + C_2 x_2 + C_3 x_3 = 0. \tag{2.21}$$

This shows that the trajectory of the particle lies in the plane (2.21). Taking this plane as a coordinate plane, without any loss of generality we can set

$$x_3 = 0, \quad \dot{x}_3 = 0.$$

It is convenient to study our plane motion in polar coordinates, which are obtained from spherical coordinates as

$$\begin{aligned} x_1 &= r\sin\theta\cos\phi, \\ x_2 &= r\sin\theta\sin\phi, \\ x_3 &= r\cos\theta, \end{aligned} \tag{2.22}$$

with

$$\varphi = \frac{\pi}{2}, \quad \dot{\varphi} = 0. \tag{2.23}$$

If the Lagrangian is written in polar coordinates, one finds:

$$\left(\frac{L}{m_0}\right)^2 = c^2\left(1-\frac{n}{r}\right) - \left(1-\frac{n}{r}\right)^{-1}\left(\frac{dr}{dt}\right)^2 - r^2\left(\frac{d\theta}{dt}\right)^2. \tag{2.24}$$

Since Lagrangian (2.24) does not explicitly depend on time t and angle θ, the associated first integrals of the motion are

$$\dot{r}\frac{\partial L}{\partial \dot{r}} + \dot{\theta}\frac{\partial L}{\partial \dot{\theta}} - L = const, \tag{2.25}$$

and

$$\frac{\partial L}{\partial \dot{\theta}} = const, \tag{2.26}$$

representing the energy and areas first integrals. Supposing that the proper time τ is considered as parameter on geodesic [20], that is

$$ds = c d\tau, \tag{2.27}$$

the first integrals (2.25) and (2.26) can explicitly be written as

$$c^2\left(1-\frac{n}{r}\right)\frac{dt}{dr} = \frac{E}{m_0}, \tag{2.28}$$

$$r^2\frac{d\theta}{d\tau} = \frac{M}{m_0}, \tag{2.29}$$

where E, M are constants of integration. The quantity E signifies the total energy of the probe particle, since in the non-relativistic approximation ($v \ll c$) (2.28) writes

$$E \simeq m_0c^2 + \frac{1}{2}m_0v^2 - \frac{Gm_0M_0}{r} + \dots \tag{2.30}$$

Indeed, making allowance for the relation cosmic time - proper time [8]

$$\frac{dt}{d\tau} = \left(1 - \frac{v^2}{c^2} - \frac{n}{r}\right)^{-\frac{1}{2}} \tag{2.31}$$

within the same approximation of the motion (2.31) yields

$$\frac{dt}{d\tau} \simeq 1 + \frac{1}{2}\frac{v^2}{c^2} + \frac{n}{r} \tag{2.32}$$

and (2.28) goes to (2.30).

The constant quantity M appearing in (2.29) can be interpreted as the angular momentum of the probe particle and, if the proper time τ is considered, the areas law is satisfied. But, if the cosmic time is concerned, this law is not valid anymore. Indeed, (2.28) and (2.29) lead to

$$\frac{1}{2}r^2\frac{d\theta}{dt} = \frac{1}{2}\frac{Mc^2}{E}\left(1 - \frac{n}{r}\right), \tag{2.33}$$

which shows the dependence of M on r. We also mention that the metric (2.15) is expressed in terms of cosmic time.

The equation of trajectory is established using identity

$$c^2 = c^2\left(1 - \frac{n}{r}\right)\left(\frac{dt}{d\tau}\right)^2 - \left(1 - \frac{n}{r}\right)^{-1}\left(\frac{dr}{dt}\right)^2 - r^2\left(\frac{d\theta}{d\tau}\right)^2, \tag{2.34}$$

given by the Lagrangian (2.24) and relation (2.27). In such conjecture, eliminating the time-dependence by means of (2.28) and (2.29), one obtains

$$\frac{1}{r^4}\left(\frac{dr}{d\tau}\right)^2 = \left[\frac{n}{r^3} - \frac{1}{r^2} + \left(\frac{m_0 c}{M}\right)^2\frac{n}{r} - \left(\frac{m_0 c}{M}\right)^2\left(1 - \frac{E^2}{m_0^2 c^4}\right)\right], \tag{2.35}$$

or, if substitution $z = n/r$ is used,

$$\left(\frac{dz}{d\tau}\right)^2 \equiv P(z) = z^3 - z^2 + \left(\frac{nm_0 c}{M}\right)^2 z - \left(\frac{nm_0 c}{M}\right)^2\left(1 - \frac{E^2}{m_0^2 c^4}\right), \tag{2.36}$$

and finally, by integration [5],

$$\theta = \pm\int\frac{dz}{\sqrt{P(z)}} + const., \tag{2.37}$$

where the sign $(+)/(-)$ is taken for increasing/decreasing z, respectively. Since $P(z)$ is a third degree polynomial, the integral (2.37) is an elliptic integral of the first kind, and its inverse is an elliptic function ([6]).

2.5 Planetary perihelion displacement

For

$$M \geq \sqrt{3} n m_0 c, \tag{2.38}$$

and

$$\begin{aligned} & m_0 c^2 \left\{ 1 + \frac{2}{3} \left(\frac{M}{2 n m_0 c} \right)^2 \left[1 - \frac{1}{2} \left(\frac{3 n m_0 c}{M} \right)^2 - \left(1 - 3 \frac{n^2 m_0^2 c^2}{M^2} \right)^{3/2} \right] \right\}^{1/2} \\ & \leq E \leq m_0 c^2 \left\{ 1 + \frac{2}{3} \left(\frac{M}{3 n m_0 c} \right)^2 \right. \\ & \left. \times \left[1 - \frac{1}{2} \left(\frac{3 n m_0 c}{M} \right)^2 + \left(1 - 3 \frac{n^2 m_0^2 c^2}{M^2} \right)^{3/2} \right] \right\}^{1/2} \end{aligned} \tag{2.39}$$

the roots of polynomial $P(z)$ can be written in trigonometric form [5]:

$$\begin{aligned} z_1 &= -\frac{2}{3} \left[1 - 3 \left(\frac{n m_0 c}{M} \right)^2 \right]^{1/2} \cos \omega + \frac{1}{3}, \\ z_2 &= \frac{2}{3} \left[1 - 3 \left(\frac{n m_0 c}{M} \right)^2 \right]^{1/2} \cos \left(\omega + \frac{\pi}{3} \right) + \frac{1}{3}, \\ z_3 &= \frac{2}{3} \left[1 - 3 \left(\frac{n m_0 c}{M} \right)^2 \right]^{1/2} \cos \left(\omega - \frac{\pi}{3} \right) + \frac{1}{3}, \end{aligned} \tag{2.40}$$

with

$$0 \leq \omega \leq \frac{\pi}{3} \tag{2.41}$$

and

$$\cos 3\omega = \frac{-1 + \frac{1}{2} \left(\frac{3 n m_0 c}{M} \right)^2 - \frac{3}{2} \left(\frac{3 n m_0 c}{M} \right)^2 \left(1 - \frac{E^2}{m_0^2 c^4} \right)}{\left[1 - 3 \left(\frac{n m_0 c}{M} \right)^2 \right]^{3/2}}. \tag{2.42}$$

The root z_1 on interval (2.41) is negative and has no physical significance. Nevertheless, it cannot be ignored, since results in which it is implicated have a straight interpretation; so, we shall only formally operate with it.

Under these circumstances, the change of variable

$$z = -\frac{2}{3}\left[1 - 3\left(\frac{nm_0c}{M}\right)^2\right]^{1/2} + \frac{1}{3}$$
$$+ \frac{2}{\sqrt{3}}\left[1 - 3\left(\frac{nm_0c}{M}\right)^2\right]^{1/2} \cos\left(\omega + \frac{\pi}{6}\right)\sin^2\varphi, \qquad (2.43)$$

transforms (2.37) into a complete elliptic integral of the first kind ([6])

$$\theta = \pm 2\gamma^{-1}\int \frac{d\varphi}{\sqrt{1 - k^2\sin^2\varphi}} + const. \qquad (2.44)$$

of modulus

$$k^2 = \frac{\sqrt{3} - tan\omega}{\sqrt{3} + tan\omega} \qquad (2.45)$$

where

$$\gamma^{-1} = \left\{2 \cdot 3^{-1/2}\left[1 - 3\left(\frac{nm_0c}{M}\right)^2\right]^{1/2} \cos\left(\omega - \frac{\pi}{5}\right)\right\}^{-1/2}. \qquad (2.46)$$

For an initial value

$$z_1 \leq z_0 \leq z_2, \qquad (2.47)$$

relation (2.43) allows limitations of the possible formal motions to the interval

$$z \in [z_1, z_2] \qquad (2.48)$$

which corresponds to planetary motions. Since in this interval the variable z increases, one considers only the sign (+) in integral (2.44), in order to calculate the perihelion displacement of the planets.

Let us now settle the integration limits in (2.44). To this end, one observes in (2.43) that the maximum value on the trajectory "radius" corresponds to $\varphi = 0$, while the minimum value of the "radius" of the same trajectory is obtained for $\varphi = \pi/2$. One also observes that the minimum value of the trajectory "radius" corresponds to both $\varphi = -\pi/2$ and $\varphi = +\pi/2$. Therefore the quantity

$$\delta\omega = 4\gamma^{-1}\int_0^{\pi/2} \frac{d\varphi}{\sqrt{1 - k^2\sin^2\varphi}} - 2\pi \qquad (2.49)$$

gives the same variation for the polar angle θ when the probe body covers once the trajectory, starting from the minimum value of the trajectory "radius" and coming back to it. If we agree to call aphelion the "maximum radius" of trajectory, and perihelion the minimum one, the quantity (1.49) shall express the exact value of the perihelion displacement at the end of a complete rotation.

2.6 The motion trajectory. Motion degeneration

The integral (2.44) allows one to establish the analytical expression for the planetary trajectory. First, one observes that the change of variable

$$s = \sin\varphi \tag{2.50}$$

leads to Legendre form of the integral ([6])

$$\theta = 2\gamma^{-1}\int \frac{ds}{\sqrt{(1-s^2)(1-k^2s^2)}} + const. \tag{2.51}$$

Its inverse is Jacobi's elliptic function sn [6]

$$s = sn\left(\frac{1}{2}\gamma\theta + const.\right) \tag{2.52}$$

with periods

$$\omega_1 = 4K\gamma^{-1} \tag{2.53}$$

and

$$\omega_2 = 2K'\gamma^{-1}, \tag{2.54}$$

where

$$\begin{aligned} K &= \int_0^{\pi/2} \frac{d\varphi}{\sqrt{1-k^2\sin^2\varphi}}, \\ K' &= i\int_0^{\pi/2} \frac{d\varphi}{\sqrt{1-k'^2\sin^2\varphi}}, \\ k^2 &+ k'^2 = 1, \quad i = \sqrt{-1}. \end{aligned} \tag{2.55}$$

The equation of planetary trajectory can be found by introducing (2.52) into (2.43). The result is:

$$\begin{aligned} z = &-\frac{2}{3}\left[1-3\left(\frac{nm_0c}{M}\right)\right]^{1/2}\cos\omega + \frac{1}{3} \\ &+ \frac{2}{\sqrt{3}}\left[1-3\left(\frac{nm_0c}{M}\right)^2\right]^{1/2}\cos\left(\omega+\frac{\pi}{6}\right) sn^2\left(\frac{1}{2}\gamma\theta + const.\right). \end{aligned} \tag{2.56}$$

As we shall later show, this formula generalizes the classical ellipse trajectory.

Admitting that the geometric significance of the radial coordinate is well-defined by classical Keplerian motions, let us now use (2.56) to study these cases. Since r is an elliptic function, they can only be studied by using the degenerations of (2.52); therefore, it is necessary to establish conditions for which the periods (2.53) and (2.54) of the elliptic function (2.52) become infinite.

In view of dependence of periods ω_1 and ω_2 on moduli k and k' of the elliptic integral (2.51), as shown by (2.55), we shall have either

$$k^2 \to 0, \tag{2.57}$$

or

$$k'^2 \to 0, \tag{2.58}$$

which implies by means of (2.41)

$$\omega \to \frac{\pi}{3}, \tag{2.59}$$

respectively

$$\omega \to 0. \tag{2.60}$$

In the first case, (2.53), (2.54), and (2.52) become [5]

$$\omega_1 \to 2\pi\gamma^{-1}, \tag{2.61}$$

$$\omega_2 \to i\infty, \tag{2.62}$$

$$sn \to \sin, \tag{2.63}$$

while in the second we are left with

$$\omega_1 \to \infty, \tag{2.64}$$

$$\omega_2 \to i\pi\gamma^{-1}, \tag{2.65}$$

$$sn \to \tan h. \tag{2.66}$$

There are several possibilities to perform the above discussed degenaracies. For example, (2.57) proves to be reductible to condition

$$E \to m_0c^2 \left\{1 + \frac{2}{3}\left(\frac{M}{nm_0c}\right)^2 \left[1 - \frac{1}{2}\left(\frac{3nm_0c}{M}\right)^2 - \left(1 - 3\frac{n^2m_0^2c^2}{M}\right)^{3/2}\right]\right\}. \tag{2.67}$$

In this situation, (2.55) reduces to the precessing ellipse

$$z \to \frac{n[1 + e\cos(\nu\theta)]}{p} \tag{2.68}$$

with the parameters

$$\begin{aligned} p &\to \left.\frac{2n}{z_1 + z_2}\right|_{\omega \to \frac{\pi}{3}}, \\ e &\to \left.\frac{z_2 - z_1}{z_1 + z_1}\right|_{\omega \to \frac{\pi}{3}} \end{aligned} \tag{2.69}$$

while (2.49) goes to

$$\delta\omega \to 4\left[1 - 3\left(\frac{nm_0c}{M}\right)^2\right]^{-1/4} \int_0^{\pi/2} d\varphi - 2\pi. \tag{2.70}$$

Since $\frac{nm_0c}{M} \leq 1$, the radicals appearing in (2.67) and (2.70) can be expand into power series in this small quantity, which gives:

$$E \to m_0c^2 \left[1 - \frac{1}{2}\left(\frac{nm_0c}{M}\right)^2\right], \tag{2.71}$$

and, respectively

$$\delta\omega \to \frac{6\pi}{4}\left(\frac{nm_0c}{M}\right)^2 = \frac{6\pi G^2 M_0^2 m_0^2 c^2}{M^2}, \tag{2.72}$$

Formula (2.72) gives the planetary perihelion displacement; this result coincides with that usually given by general relativity [25]. The fact that it has been obtained as an asymptotic limit in the imaginary period ($\omega_2 \to \infty$) allows us to conclude that the real period ω_1 of the elliptic function (2.52) is different from 2π, and the precessing ellipse is only the first approximation of the planetary trajectory, since geometrical reasons rather than physical ones are taken into account.

Going back to (2.71), one observes that expression for energy differs from the Einsteinian formula, since some characteristics of gravitational interaction (the source structure through the medium of gravitational radius, the angular momentum of the probe body, etc.) are considered. This could give the idea of introducing these characteristics in the momentum definition in a straight way, as invariantive mechanics [16] does. But our result is a direct consequence of utilizing elliptic functions. The fact that the energy of the probe body depends on the inverse of angular momentum shows that, far from the field source, it becomes independent of M. Therefore, one can expect to obtain Einstein's relation in case of a free body. Indeed, if $M \to \infty$ in (2.71), we are left with

$$E \to m_0 c^2. \tag{2.73}$$

It is interesting to observe a certain analogy between relation (2.71) and that given by non-linear electrodynamics. Indeed, if the gravitational radius n is replaced by the effective electron radius n_e, that is

$$n \to n_e \approx \frac{e^2}{4\pi\varepsilon_0 m_0 c^2}, \tag{2.74}$$

where e, ε_0 are the electron charge and vacuum permittivity, respectively, we have

$$E = m_0 c^2 \left[1 - \frac{1}{2}\left(\frac{n_e m_0 c}{M}\right)^2\right], \tag{2.75}$$

which is identical to that given in [10] for the first order of approximation of parameter $(\frac{n_\varepsilon m_0 c}{M})$. Consequently, the gravitational interaction has to be a non-linear interaction.

Degeneration for which

$$n \to 0 \tag{2.76}$$

is also a degeneration of the type (2.57), that is

$$\omega_1 \to 2\pi \tag{2.77}$$

in which case the equation (2.56) of the planetary trajectory reduces to the Newtonian ellipse

$$\frac{1}{r} \to \frac{1 + e_0 \cos(\theta + const.)}{p_0} \tag{2.78}$$

with the parameters [8]

$$\begin{aligned} p_0 &= \left(\frac{M'}{m_0}\right)^2 \frac{1}{GM_0}, \\ e_0 &= \left[1 - \frac{E'}{m_0}\left(\frac{M'}{m_0}\right)^2 \frac{1}{G^2 M_0^2}\right]^{1/2}, \end{aligned} \tag{2.79}$$

where E' and M' are the classical values of energy and angular momentum.

The condition through which the degeneracy (2.58) can be performed writes

$$E \to m_0c^2\left\{1+\frac{2}{3}\left(\frac{M}{3nm_0c}\right)^2\left[1-\frac{1}{2}\left(\frac{3nm_0c}{M}\right)^2\right.\right.$$
$$\left.\left.+\left(1-3\frac{n^2m_0^2c^2}{M^2}\right)^{3/2}\right]\right\}^{1/2}. \tag{2.80}$$

In this case, the trajectory equation becomes

$$\frac{1}{r} \to \frac{1}{n}\left\{-\frac{2}{3}\left[1-3\left(\frac{nm_0c}{M}\right)^2\right]^{1/2}+\frac{1}{3}\right.$$
$$\left.+\,tanh^2\left[\left(1-3\left(\frac{nm_0c}{M}\right)^2\right)\frac{\theta}{2}+const.\right]\right\} \tag{2.81}$$

Study of degeneration of the elliptic function (2.52) is particularly interesting. It allows introduction of a finitude criterion for the planetary trajectories, in terms of the periods of the function. This criterion can be stated as follows: the motion trajectories (2.56) are finite in the asymptotically infinite limit of the imaginary period (2.54); on the contrary, they become infinite in the asymptotically infinite limit of the real period (1.53).

Indeed, if M increases from value $\sqrt{3}nm_0$ to infinity, the energy given by (2.67) decreases from $\sqrt{\frac{8}{9}}m_0c^2$ to m_0c^2 so that the trajectories (2.56) are finite, but for the same variation of M the energy given by (2.80) increases from $\sqrt{3}nm_0c$ to infinity, which determines non-finitude of the planetary orbits. Such a result was also obtained by Kaplan [13].

To conclude, we considered the free motion of a particle for an eventual comparison with the classical Kepler problem, where the radial coordinate still has the significance of the radial distance from the origin. Our analysis shows that this case is obtained for $n \to 0$, which means that the mass of the field source is negligible. If it is precisely zero, we meet the Newtonian kinematics conditions corresponding to Euclidean geometry. Therefore, using elliptic functions, the radial coordinate can still be defined as a continuous quantity, having significance of a radius vector, as in the Euclidean space, but only in the limits of nullity of Schwarzschild universe of the metric. In general, for $n \neq 0$, the radial coordinate can have the usual geometric significance only if conditions (2.67) and (2.80) still allow circular (or almost circular) orbits. But in this case the "radius vector" has to be redefined under a complex form, the real part giving the average of the quantity, and the imaginary part its dispersion. This assertion is based on the idea that any quantity which cannot be exactly measured can be represented by a complex number ([2],[3]).

2.7 Null geodesics. Motion degeneracy

Since for the light beam [25]

$$d\tau \to 0, \tag{2.82}$$

the motion constants E and M met in (2.28) and (2.29) become infinite, but their ratio given by

$$z^{-2}(1-z)^{-1}\frac{d\theta}{dt} = \lim\left(\frac{c}{n}\right)^2\frac{M}{E} \to \frac{c}{n^2}D \tag{2.83}$$

remains finite. Here D is a constant which has a dimension of length.

The differential equation of the light beam is obtained by means of relation

$$c^2(1-z)-(1-z)^{-1}\left(\frac{n}{z^2}\right)^2\left(\frac{dz}{d\theta}\right)^2\left(\frac{d\theta}{dt}\right)^2-\left(\frac{n}{z^2}\right)^2\left(\frac{d\theta}{dt}\right) = 0, \tag{2.84}$$

given by the form of the Lagrangian (2.24) with condition (2.82) and relation (2.83) by eliminating any time-dependence. One finds

$$\left(\frac{dz}{d\theta}\right)^2 = z^3 - z^2 + \frac{n^2}{D^2} \tag{2.85}$$

and, by integration

$$\theta = \pm\int\frac{dz}{\sqrt{z^3 - z^2 + \frac{n^2}{D^2}}} + const. \tag{2.86}$$

If (r,θ) are interpreted as polar coordinates of the Euclidean plane, the constant D of (2.86) can be interpreted as the impact parameter, that is the length of the perpendicular from the coordinate origin to the asymptote to trajectory [8]. Indeed, the formulas of the plane Euclidean geometry give the following expression for the perpendicular drawn from the coordinate origin to the tangent of a given curve, in polar coordinates [8]

$$d = \frac{r}{\sqrt{1 + \frac{1}{r^2}\left(\frac{dt}{d\theta}\right)^2}} = \frac{n}{\sqrt{z^2 + \left(\frac{dz}{d\theta}\right)^2}}. \tag{2.87}$$

Asymptote is the tangent at infinity ($z \to 0$); in this case, the "impact parameter" is the value of d for $z \to 0$. This value is obtained from (2.87) and (2.85), being equal to D.

Since the roots of the cubic from (2.85) can be written in trigonometric form

$$\begin{aligned} z_1 &= -\frac{2}{3}\cos\omega + \frac{1}{3}, \\ z_2 &= \frac{2}{3}\cos\left(\omega + \frac{\pi}{3}\right)^2 + \frac{1}{3}, \\ z_3 &= \frac{2}{3}\cos\left(\omega - \frac{\pi}{3}\right) + \frac{1}{3} \end{aligned} \tag{2.88}$$

with

$$\cos 3\omega = -1 + \frac{3}{2}\left(\frac{3n}{D}\right)^2, \tag{2.89}$$

the successive change of variable

$$z = -\frac{2}{3}\cos\omega + \frac{1}{3} + \frac{2}{\sqrt{3}}\cos\left(\omega + \frac{\pi}{6}\right)\sin^2\varphi \tag{2.90}$$

and

$$s = \sin\varphi \tag{2.91}$$

brings (2.86) to Legendre form of elliptic integral of the first kind ([6])

$$\theta = \frac{2}{\sqrt{\frac{2}{\sqrt{3}}\cos\left(\omega - \frac{\pi}{6}\right)}} \int \frac{ds}{\sqrt{(1-s^2)(1-k^2s^2)}}, \tag{2.92}$$

of the same modulus (2.45). The inverse of integral (2.92) is Jacobis elliptic function

$$s = sn\left(\frac{\gamma'}{2}\theta + const.\right), \tag{2.93}$$

where

$$\gamma' = [2 \cdot 3^{-1/2} \cos\left(\omega - \frac{\pi}{6}\right)]^{1/2}, \tag{2.94}$$

with periods of the same form as (2.53), (2.54).

The equation of trajectory of the light beam is found by introducing (2.93) into (2.90). The result is:

$$z \to -\frac{2}{3}\cos\omega + \frac{1}{3} + \frac{2}{\sqrt{3}}\cos\left(\omega + \frac{\pi}{6}\right) sn^2\left(\frac{1}{2}\gamma'\theta + const.\right).$$

Since the trajectory of motion is expressed by an elliptic function, it is interesting to study its degeneracies. So, if the real period tends to infinity, that is

$$n \to 0,$$

the trajectory of motion is the straight line

$$\frac{1}{r} \to \frac{1}{D}\cos(\theta + const.) \tag{2.95}$$

If the imaginary period is infinite, which means

$$D \to \frac{3\sqrt{3}}{2}n, \tag{2.96}$$

the trajectory becomes

$$\frac{1}{r} \to \frac{1}{n}\left[-\frac{1}{3} + tanh^2\left(\frac{1}{2}\theta + const.\right)\right]. \tag{2.97}$$

This result shows that, for values smaller than (2.96) of the "impact parameter", the light beam "falls" on the source body. Using a different procedure, Zeldovich and Novikov [13] obtain the same result.

2.8 The Fock metric. Null geodesics and radius significance

The Fock metric is given by

$$ds^2 = c^2\frac{2R-n}{2R+n}dt^2 - \frac{2R+n}{2R-n}dR^2 - \frac{1}{4}(2R+n)^2(d\theta'^2 + \sin^2\theta' d\varphi'^2), \tag{2.98}$$

where (R, θ', φ') are "harmonic coordinates". We shall study only null geodesics for this metric, in order to compare them with those associated with Schwarzschild metric (2.15).

Observations developed in the previous paragraph of this chapter can be also applied to metric (2.98), written in polar coordinates so that the Lagrangian writes

$$\left(\frac{L}{m_0}\right)^2 = c^2\frac{2R-n}{2R+n} - \frac{2R+n}{2R-n}\dot{R}^2 - \frac{1}{4}(2R+n)^2\dot{\theta}'^2, \tag{2.99}$$

while the motion first integrals are [8]

$$\frac{2R-n}{2R+n}\frac{dt}{d\tau} = \varepsilon_0, \tag{2.100}$$

and

$$\frac{1}{4}(2R+n)^2\frac{d\theta'}{d\tau} = \mu, \tag{2.101}$$

where ε_0 and μ are constants of integration.

If

$$\varepsilon_0 = 1 + \frac{E}{c^2}, \tag{2.102}$$

where E is a new constant, then (2.100), in view of (2.31), becomes

$$E \simeq \frac{1}{2}v^2 - \frac{GM_0}{r}, \tag{2.103}$$

which means that E signifies the total specific energy of the probe particle.

The quantity μ can be interpreted as the angular momentum of the probe particle. Independently on the approached time (proper or cosmic), the areas law cannot be satisfied by any condition. There is, nevertheless, a situation in which this law is satisfied, namely if the origin of the radial distances from the field source as being considered $\frac{n}{2}$, instead of zero; in this case, obviously, (2.101) stands for the equivalent of the areas law.

Since the road traveled by light beam in the gravitational field of the source-body is a geodesic of length zero relation (2.82) is valid. For such a geodesic parameter τ cannot be used, but we can suppose that it is the limit of the geodesics itselves for the condition (2.82) so that, starting from (2.100) and (2.101), one obtains:

$$\frac{1}{4}\frac{(2R+n)^3}{2R-n}\frac{d\theta'}{dt} = \frac{\mu}{\varepsilon_0} = cD'. \tag{2.104}$$

This relation remains valid in the limit (2.82). The constant D' has dimension of a length and, similarly to procedure used in (2.82), it can be shown that it signifies the "impact parameter".

The differential equation of the light beam trajectory is obtained by means of identity

$$c^2\frac{2R-n}{2R+n}(dt)^2 - \frac{2R+n}{2R-n}(dR)^2 - \frac{1}{4}(2R+n)^2(d\theta')^2 = 0, \quad (2.105)$$

given by the Lagrangian (2.99) and condition (2.82). Using substitutions

$$u = \frac{R}{D'}, \quad \alpha = \frac{n}{2D'}, \quad (2.106)$$

we still have

$$c^2\left(\frac{u-\alpha}{u+\alpha}\right) - D'^2\left(\frac{u-\alpha}{u+\alpha}\right)\left(\frac{du}{d\theta'}\right)^2\left(\frac{d\theta'}{dt}\right)^2 - D'^2(u+\alpha)^2\left(\frac{d\theta'}{dt}\right)^2 = 0. \quad (2.107)$$

Eliminating the time dependence by means of (2.104), one finally finds [8]

$$\left(\frac{du}{d\theta'}\right)^2 = P(u) = (u+\alpha)^4 - (u+\alpha)(u-\alpha). \quad (2.108)$$

Since $P(u)$ is a fourth degree polynomial, θ' is expressed in terms of R as an elliptic integral of the first kind ([6])

$$\theta' = \pm\int\frac{du}{\sqrt{P(u)}} + const. \quad (2.109)$$

and, reciprocally, R is an elliptic function of θ'.

To write the elliptic function, let us first determine the roots of the polynomial $P(u)$. A simple - but somewhat long - calculation gives [4]

$$\begin{aligned} u_1 &= -\frac{2}{\sqrt{3}}\cos\omega - \alpha, \\ u_2 &= -\alpha, \\ u_3 &= \frac{2}{\sqrt{3}}\cos\left(\omega + \frac{\pi}{3}\right) - \alpha, \\ u_4 &= \frac{2}{\sqrt{3}}\cos\left(\omega - \frac{\pi}{3}\right) - \alpha, \end{aligned} \quad (2.110)$$

where

$$\cos 3\omega = 3\sqrt{3}\alpha \tag{2.111}$$

together with the condition

$$3\sqrt{3}\alpha \leq 1. \tag{2.112}$$

Using these roots, a successive change of variables

$$z^2 = \frac{u - u_1}{u - u_2}, \tag{2.113}$$

and

$$s = z\left(\frac{u_2 - u_4}{u_1 - u_4}\right)^{1/2}, \tag{2.114}$$

bring integral (2.109) to the Legendre form of the elliptic integral of the first kind

$$\theta' = \pm 2(\gamma'')^{-1} \int \frac{ds}{\sqrt{(1 - s^2)(1 - k^2 s^2)}} + const. \tag{2.115}$$

of modulus

$$k^2 = \frac{(u_1 - u_4)(u_2 - u_3)}{(u_1 - u_3)(u_2 - u_4)}, \tag{2.116}$$

where we denoted

$$\gamma'' = [(u_2 - u_4)(u_1 - u_3)]^{1/2}. \tag{2.117}$$

Explicitly, (2.116) and (2.117) are given by

$$k^2 = \frac{\cos(\omega + \frac{\pi}{3})\cos(\omega - \frac{\pi}{6})}{\cos(\omega - \frac{\pi}{3})\cos(\omega + \frac{\pi}{6})}, \tag{2.118}$$

and

$$\gamma'' = \left[\frac{4}{\sqrt{3}}\cos\left(\omega - \frac{\pi}{3}\right)\cos\left(\omega + \frac{\pi}{6}\right)\right]. \tag{2.119}$$

The inverse of integral (2.115) is Jacobi's *sn* function

$$s = sn\left(\gamma''\frac{\theta}{2} + const.\right), \tag{2.120}$$

its periods being of the same form (2.53) and (2.54), but moduli given by (2.118) and $k' = \sqrt{1-k^2}$. Equation of trajectory is then obtained by means of (2.113), (2.114) and (2.120) and has the form:

$$u = \frac{1 - 2\cos\frac{\pi}{6}\cos(\omega - \frac{\pi}{6})\cos^{-1}(\omega - \frac{\pi}{3})}{\alpha - \frac{2}{\sqrt{3}}\cos\omega + 2\alpha\cos\frac{\pi}{6}\cos(\omega - \frac{\pi}{6})}$$
$$\times \frac{sn^2(\gamma''\frac{\theta'}{2} + const.)}{\cos^{-1}(\omega - \frac{\pi}{3})sn^2(\gamma''\frac{\theta'}{2} + const.)}. \tag{2.121}$$

Let us now analyse, in case of this metric, situations in which degeneration of the elliptic function (2.120) can be obtained. Degeneration (2.57) is obtained through condition (2.77). In this circumstance, integral (2.115) leads to the straight line

$$\frac{1}{R} \to \frac{\cos(\theta' + const.)}{D'}. \tag{2.122}$$

The condition for degeneration (2.58) is (2.96). In this case, integral (2.115) takes the form

$$\theta' = \pm 2\int \frac{ds}{(1 - s^3)}. \tag{2.123}$$

Using the change of variable

$$s = \sqrt{\left(\frac{u - u_1}{u - u_2}\right)\left(\frac{u_2 - u_4}{u_1 - u_4}\right)} \tag{2.124}$$

one finds

$$\frac{u - u_1}{u - u_2} = \frac{u_2 - u_4}{u_1 - u_4} tanh^2\frac{\theta'}{2}. \tag{2.125}$$

Finally, the equation of trajectory is found as

$$R \to n\frac{tanh^2\frac{\theta'}{2} - 21}{3 - tanh^2\frac{\theta'}{2}}. \tag{2.126}$$

Therefore, even if the degeneration conditions of the elliptic functions (2.120) and (2.93) are the same, the trajectories for the Schwarzschild and Fock metrics are different. The explanation lies in the fact that the coordinate systems (r, θ, φ) and (R, θ', φ') appearing in the two metrics do not have the same physical significance.

Let us admit, nevertheless, that the angular variables (θ, φ) and (θ', φ') have the same significance. Then the elliptic functions (2.120) and (2.93) are equivalent ([6]) and between them exists at most an algebraic relation of the form

$$r = R + \frac{n}{2}. \tag{2.127}$$

Indeed, the integral (2.86), expressed in terms of $\sigma = \frac{1}{r}$ and written in terms of a new variable

$$\sigma = \sigma' + \frac{1}{3n}, \tag{2.128}$$

becomes

$$\theta = \pm\frac{2}{n}\int \frac{d\sigma'}{\sqrt{4\sigma'^2 - g_2'\sigma' - g_3'}}, \tag{2.129}$$

with

$$g_2' = \frac{4}{3n^2}, \; g_3' = \frac{4}{nD^2} + \frac{8}{27n^3}, \tag{2.130}$$

whose inverse is the Weierstrass elliptic function Γ

$$\sigma' = \Gamma\left(\frac{n\theta}{2} + const.\right). \tag{2.131}$$

If in integral (2.115), written in terms of R, one performs the consecutive change of variables

$$R = -\frac{n}{2} + \frac{1}{t}, \tag{2.132}$$

and

$$t = t' + \frac{1}{3n}, \tag{2.133}$$

its form becomes similar to (2.129), but in variable t'. The inverse of this integral also is the Weierstrass elliptic function

$$t' = \Gamma\left(\frac{n\theta'}{2} + const.\right). \tag{2.134}$$

Since the elliptic functions (2.131) and (2.134) are equivalent (for details see [6]), the algebraic relation between them writes

$$\sigma = t', \tag{2.135}$$

and so we obtain (2.127). This result is identical to that given in paper [8]. Therefore, we can keep the usual, Euclidean, significance of angles, but the definition of radius vector is a matter of choice.

There are various procedures to remove indeterminacy of the "radius vector"; the most frequently used is to consider Lorentz transformation as being valid in spherical coordinates, as it shall be seen in the following paragraph.

2.9 Validity of the Lorentz transformation in spherical coordinates

The metrics with spherical symmetry - and, generally, metrics demanding spatial symmetry - have deep roots in human beings way of thinking. Indeed, we cannot accept other way of getting information from Universe than a null radial geodesic. In relativistic investigations the angular coordinates (θ, φ) can keep their usual, Euclidean, significance, but the radial coordinate is submitted to a wide arbitrariety, due to impossibility of direct determination of astronomical distances. Various coordinate systems (harmonic, isotropic, etc.) have bee imagined to replace this kind of indecision. In this case, it is also necessary to study the behavior of the null radial geodesics, to realize in what amount they keep their significance. Such an investigation implies analysis of validity of the Lorentz transformations in spherical coordinates. The result of this study [17] shows the existence of a much larger symmetry group containing as a subgroup the ortochron improper Lorentz group, when the radial coordinate tends to infinity. More precisely, transition from the usual coordinate r to R through the Lorentz group takes place if

$$\lim \left(\frac{R}{r}\right) = K(\theta, \varphi), \tag{2.136}$$

showing that the limit of their ratio affects the conform metric on the unit sphere. More than that, there must exist a connection between the two parts of the metric, taken as separate entities, the space-time (r, t) and the angular (θ, φ) parts: transformation on unit sphere induces certain transformations in space-time metric, and reciprocally. We shall come back later on this matter, showing that this interdependence leads to some already known results.

Applying the previous analysis to metrics Schwarzschild (2.15) and Fock (2.98), it follows that the conformity factor of angular metrics is always unitary; this is an expected result, since these metrics are asymptotically flat. Therefore, the absence of interactions selects for the studied metrics one and the same variety - the unit sphere. If the frame is defined as spatial directions, transformation which performs transition from Schwarzschild to Fock metrics can be interpreted as a coordinate transformation within the same frame, i.e. the frame with origin at the source body. Indeed, denoting by r and R the Schwarzschild and Fock "radius vectors", respectively, the coordinate change (2.127) transforms the Schwarzschild's arc element (2.15) into Fock's arc element (2.98).

The coordinate transformations do not affect the nature of variety, but only its points; the direction changes transform the variety into itself. This shows that supplementary precautions have to be taken in order to decide which transformations refer to coordinates, and which to the frame.

2.10 Singularities in spherical metrics. The Kruskal coordinates and the hidden symmetries. Perspectives

Metric (2.15) becomes singular for

$$r = n, \tag{2.137}$$

i.e. for Schwarzschild radius. But the gravitational field presents such a singularity for no known objects in the Universe. It can appear only in case of vacuum solution of Einstein's equations

$$R_{\mu\nu} = 0, \ \mu, \nu = 0, 1, 2, 3, \tag{2.138}$$

and, consequently, it is not real if radius n is located inside the sourcebody. For example, for the Earth

$$n_E = 0.5cm, \tag{2.139}$$

and for the proton

$$n_p = 10^{-50} cm. \tag{2.140}$$

which values are much smaller than the radii of these objects.

Nevertheless, it would by interesting to accept existence of a body enough small and massive for n to be greater then its characteristic value in vacuum. The Schwarzschild solution then maintains in this case and, obviously, the above mentioned singularity appears in vacuum. Shall this singularity show itself? Within the frame of Einsteinian theory this question can be answered only from the point of view of curvature invariants. If these invariants are combined, one realizes that no jump appears for the Schwarzschild radius, even if they become singular at origin [13]. This would indicate that the "apparent singularity" for the Schwarzschild radius is only a property of the coordinate system used by us [11]. Indeed, if for this radius at least one of the curvature invariants would be singular, then, by virtue of covariance, this singularity would be present in all used coordinate systems.

Kruskal [12] found a coordinate system in which Schwarzschild singularity can be removed, if a more complicated topology of the Universe is accepted. In this respect, the new coordinates u and v are chosen in such a way so as to satisfy the following equations

$$u^2 - v^2 = (cT)^2 \left(\frac{r}{n} - 1\right) \exp \frac{r}{n} \tag{2.141}$$

and

$$\frac{2uv}{u^2 + v^2} = tan\, h \left(\frac{ct}{n}\right), \tag{2.142}$$

where T is some constant. The metric (1.15) then takes the form

$$ds^2 = \frac{4n^3}{r(cT)^2} \exp \left(-\frac{r}{n}\right) (dv^2 - du^2) - r^2 d\Omega^2. \tag{2.143}$$

According to (2.141), r is now a function of $(u^2 - v^2)$. The metric is non-singular, under condition that r^2 is positive definite and does not suffer jump, that is

$$u^2 r^2 > v^2 - (cT)^2. \tag{2.144}$$

As a result, for the time interval

$$0 < v < cT \tag{2.145}$$

and all u real, the metric is a smooth, finite function of u. Indeed, even $g_{\theta\theta}$ and $g_{\varphi\varphi}$ do not cancel for $u = 0$, so that, when passing to the coordinate origin $u \to 0$, nothing stops us to go further to negative values of r. Consequently, the space described by (2.143) is free from singularities and consists of two identical bands with $u > 0$ and $u < 0$, which continuously pass one into the other at the ramification point $u = 0$. The bands separate for $v = cT$ and the metric presents a real singularity at

$$u = \pm\sqrt{v^2 - c^2T^2}, \tag{2.146}$$

corresponding to $r = 0$. Nevertheless, even in this case, there are no singularities in the metric for $u = v$, which would correspond to Schwarzschild radius n.

It is really interesting to observe that the Kruskal transformation mechanism puts into evidence the fact that removal of singularities is performed through the parameters

$$\begin{aligned} X &= (cT)^2 \left(\frac{r}{n} - 1\right) \exp\left(\frac{r}{n}\right), \\ Y &= (cT)^2 \sin h \left(\frac{ct}{n}\right) \left(\frac{r}{n} - 1\right) \exp\left(\frac{r}{n}\right), \\ Z &= (cT)^2 \cos h \left(\frac{ct}{n}\right) \left(\frac{r}{n} - 1\right) \exp\left(\frac{r}{n}\right). \end{aligned} \tag{2.147}$$

satisfying the identity

$$X^2 + Y^2 - Z^2 = 0. \tag{2.148}$$

They are essential in analytical representation of the spin [18] and, as we shall see, play an essential role in fixing the significance of the employed coordinates. Their general handling would correspond to certain conditions connected to the explicitation of coordinates and their presence in Kruskal transformations, as well as to existence of certain internal symmetries of the field equations, in view of their quantum significance.

Let us explicitate a such internal symmetry. In this purpose, let us first note that through (2.148) written in the form:

$$\mu^2 + \nu^2 = 1, \tag{2.149}$$

where

$$\mu = \frac{X}{Z} = \frac{1}{\cos\mathrm{h}(\frac{ct}{n})}, \quad \nu = \frac{Y}{Z} = \tan\mathrm{h}(\frac{ct}{n}), \tag{2.150}$$

the metric of the Lobachevsky plane can be produced as a Caylean metric of an Euclidean plane, for which the absoluteness is the circle with unit radius (2.149). Indeed, the absolute metric for the interior circle (2.149), i.e.

$$ds^2 = \frac{(1-\nu^2)(d\mu)^2 + 2\mu\nu d\mu d\nu + (1-\mu^2)(d\nu)^2}{(1-\mu^2-\nu^2)^2}, \tag{2.151}$$

can be brought to the form of Poincaré metric of the upper complex plane:

$$ds^2 = -4\frac{dh d\overline{h}}{(h-\overline{h})^2} = \frac{(du)^2 + (dv)^2}{v^2}, \tag{2.152}$$

by the following transformation of coordinates:

$$\mu = \frac{h\overline{h}-1}{h\overline{h}+1}, \nu = \frac{h+\overline{h}}{h\overline{h}+1} \leftrightarrow h = u + iv = \frac{\nu + i\sqrt{1-\mu^2-\nu^2}}{1-\mu}, \tag{2.153}$$

$$\overline{h} = u - iv.$$

Further, by means of substitutions

$$\mu = e\cos\omega, \quad \nu = e\sin\omega, \quad \varphi = \tan \mathrm{h}\psi \tag{2.154}$$

the metric (2.152) becomes

$$(ds)^2 = \left(\frac{de}{1-e^2}\right)^2 + \frac{e^2}{1-e^2}(d\omega)^2 \equiv (d\psi)^2 + \sin h^2\psi(d\omega)^2. \tag{2.155}$$

The complex parameter h from equation (2.153) has now a direct connection with the theory of classical Newtonian type potentials via a harmonic map. In order to show this relationship we write here h from Eq. (2.153) in terms of (e, ω). We have

$$h = i\frac{\cos h\chi + \sin h\chi e^{-i\omega}}{\cos h\chi - \sin h\chi e^{-i\omega}}, \quad \chi = \frac{\psi}{2}. \tag{2.156}$$

As it happens, this equation represents a harmonic map from the usual space into the Lobacevschy plane having the metric (2.152),provided χ (and therefore ψ) is a solution of Laplace equation in free space.

Indeed, the problem of harmonic correspondences between space and the hyperbolic plane is described by the stationary values of the energy functional corresponding to the metric (2.152). This is defined as the volume integral of an integrand obtained from that metric by transforming the differentials into space gradients [14]. The stationary values of energy functional therefore correspond to solutions of the Euler-Lagrange equations for a Lagrange like

$$L = -4\frac{\nabla h \nabla \overline{h}}{(h-\overline{h})^2}. \tag{2.157}$$

Then we have

$$(h-\overline{h})\nabla^2 h - 2(\nabla h)^2 = 0 \tag{2.158}$$

and its complex conjugate. Then it is easy to see, by a direct calculation, that h from equation (2.156) verifies equation (2.158) when χ is a solution of Laplace equation, and ω is arbitrary, in the sense that it does not depend on the position space. Nevertheless, it might depend on the local time of the Newtonian type dynamics, as it turns out to be the case in the particular problems related to the case of the damped harmonic oscillator [14].

Since the dynamics induced by the Schwarschild's metric are fundamentated on the cubic (2.36) it results that, in the more general case, the cubic properties could make explicit certain hidden symmetries and consequently dynamics, which on the first sight can not be evidentiated by means of usual experiments such as those from standard case (planetary perihelion displacement, light deflexion etc.). Let be then the cubic equation in the so called binomial form:

$$a_0 x^3 + 3a_1 x^2 + 3a_2 x + a_3 = 0. \tag{2.159}$$

Here we assume that the coefficients a_k are real, displaying by a_0 the possibility of adjusting them by an arbitrary factor on account of the known arbitrariness allowed by the relations between the roots and the coefficients of an algebraic equation.

The central problem related to equation (2.159) is, of course, that of finding its roots. These are for example, the eigenvalues of a matrix 3×3. There are many methods for the general solution of this problem, conveniently coping with the purpose which that solution is serving (for details see [14]). All of the methods of solution are however centered on reducing it to that of a quadratic equation and, for physics purposes, we think it is worthwhile revealing what this actually entails. The most general theory behind the procedure has been established by Sylvester (for details see the references from [14]) and amounts to putting the equation (2.159) in the form of a sum of two perfect cubes:

$$\beta_1(x-\alpha_1)^3+\beta_2(x-\alpha_2)^3=0. \tag{2.160}$$

In this case the equation can be easily solved to give

$$x_j=\frac{\alpha_1+\alpha_2\varepsilon_j k}{1+\varepsilon_j k},\ j=1,2,3, \tag{2.161}$$

where we denoted

$$k^3=\frac{\beta_2}{\beta_1},\ \varepsilon^3=1 \tag{2.162}$$

i.e. ε_j are the cubic roots of the unity:

$$\varepsilon_1=1,\ \ \varepsilon_2=\frac{-1+i\sqrt{3}}{2},\ \varepsilon_3=\frac{-1-i\sqrt{3}}{2}.$$

The problem of solving the cubic equation is thus translated into that of finding the quantities $\alpha_1,\alpha_2,\beta_1,\beta_2$ from equation (2.160) as functions of the coefficients a_0,a_1,a_2,a_3, which are usually related to physical situations. This can be done as follows: identifying the equations (2.159) and (2.160), gives the following system of equations

$$\begin{aligned}&\beta_1+\beta_2=a_0,\ \ \beta_1\alpha_1+\beta_2\alpha_2=-a_1,\\&\beta_1\alpha_1^2+\beta_2\alpha_2^2=a_2,\ \beta_1\alpha_1^3+\beta_2\alpha_2^3=-a_3.\end{aligned} \tag{2.163}$$

Now, we always may assume that α_1 and α_2 are the roots of a quadratic equation, which we write in the form:

$$b_0\alpha^2+b_1\alpha+b_2=0.$$

This equation entails the natural identities

$$\begin{aligned}&b_0\alpha_1^2+b_1\alpha_1+b_2=0,\\&b_0\alpha_2^2+b_1\alpha_2+b_2=0.\end{aligned} \tag{2.164}$$

which we try to put in relation with the system of equations (2.163). For instance, we can obtain one equation by multiplying the first of the equations (2.164) by β_1 the second one by β_2 and then adding the results. The coefficients of b_0,b_1,b_2 in the new equation are given by a_0,a_1,a_2 from (2.163). Likewise, another equation may be obtained when multiplying the first of the equations (2.164) by $\beta_1\alpha_1$, the second one by $\beta_2\alpha_2$ and then add the resulting equations. The final result for this procedure is the following system of equations for b_0, b_1, b_2:

$$\begin{aligned} a_2 b_0 - a_1 b_1 + a_0 b_2 &= 0, \\ a_3 b_0 - a_2 b_1 + a_1 b_2 &= 0. \end{aligned} \tag{2.165}$$

This system has the solution defined, up to an arbitrary factor, by the equations

$$\frac{b_0}{a_0 a_2 - a_1^2} = \frac{b_1}{a_0 a_3 - a_1 a_2} = \frac{b_2}{a_1 a_3 - a_2^2},$$

showing that α_1 and α_2 from (2.160) are the roots of a quadratic equation:

$$(a_0 a_2 - a_1^2)x^2 + (a_0 a_3 - a_1 a_2)x + (a_1 a_3 - a_2^2) = 0$$

called Hessian associated to the cubic equation (2.159). Now we must find β_1 and β_2 from (2.160). For this we can use any pair from the four equations (2.163). The result is the same up to a factor. Because we actually need only the ratio β_2/β_1, we can however use another, more direct method, having the advantage to exhibit straightforwardly the algebraic nature of β_1 and β_2. Namely, denoting the cubic from equation (2.159) by $f(x)$, and using the equation (2.160), we find

$$f(\alpha_1) = \beta_2(\alpha_1 - \alpha_2)^3, \quad f(\alpha_2) = -\beta_1(\alpha_1 - \alpha_2)^3,$$

whence the ratio between β_1 and β_2 is given by equation

$$\frac{\beta_2}{\beta_1} = -\frac{f(\alpha_1)}{f(\alpha_2)}. \tag{2.166}$$

Therefore β_1 and β_2 are of the same algebraic nature as α_1 and α_2, because the coefficients of the cubic are real. Thus, the solution of a cubic equation can, indeed, be reduced to that of a quadratic equation. Here the quadratic in question is the Hessian of the cubic. In some other methods of solution we might have some other quadratic, but it will still be in close relationship with the Hessian.

The relation between cubic and its Hessian can be summarized as follows:

i) If a cubic has as the Hessian a perfect square, then such a cubic contains the Hessian as a factor.

ii) If the Hessian of a cubic equation has distinct roots then the cubic itself has distinct roots. There are thus two cases:

ii1) If the Hessian has real roots, then the cubic itself has one real and two complex roots.

ii2) If the Hessian has complex roots, then the cubic itself has real roots

iii) If a cubic is a perfect cube, then it has a null Hessian. Reciprocally, if a cubic has a null Hessian then it is a perfect cube.

It is therefore important to introduce the distinguished quantity, playing an essential role in the theory of cubic equations. This quantity is the discriminant of the Hessian of a cubic, also called the discriminant of the cubic itself. In view of equation (2.165) it is, obviously, given by

$$\Delta = (a_0 a_3 - a_1 a_2)^2 - 4(a_0 a_2 - a_1^2)(a_1 a_3 - a_2^2) \tag{2.167}$$

and deserves this name because, like in the case of the quadratic equation, it decides the algebraic nature of the roots of cubic equation. For instance, the second of the theorems above can be directly proved just considering the form (2.161) of the roots of the cubic. The first theorem, and the third, can be proved using equation (2.167) in the special cases to which they are referring.

As one can see from the above presentation, the Hessian of a cubic is a key tool in constructing the roots of that cubic. Sometimes, in practical problems, the physical principles even allow us to know the Hessian before we know the cubic itself, and in such cases we need to figure out the corresponding cubic, more specifically to construct the roots of this cubic. One can guess that such situations currently occur in the case of nuclear matter, and this should be indeed the case if judged from a group-theoretical point of view.

The previous development of the theory of cubic equations shows that, given the roots of the Hessian only, one cannot know the corresponding cubic equation without ambiguity. In fact the equations (2.160) and (2.161) show that to a given Hessian there corresponds a one-parameter family of triplets of numbers, each one of these triplets representing a given cubic equation. This indetermination is independent of the known property of indetermination allowed by the relations between roots and coefficients. As a matter of fact, it is even deeper than the equation (2.160) shows it, in the sense that the ratio k, which by equations (2.162) and (2.166) depends only on the quantities related to the cubic equation, may hide in itself an external phase completely independent of the cubic equation - a gauge phase. This observation and the algebraic proof that follows are due to Dan Barbilian (for details see the references from [14]). In order to better grasp the nature of this problem, we use the following identity between cubic itself (f), its Hessian (H) and its Jacobian (T):

$$4H^3 = \Delta \cdot f^2 - T^2. \tag{2.168}$$

The expression in right hand side of this equation can be decomposed into two factors each of the third degree, because the cubic and its Jacobian are prime with respect to each other. On the other hand, the left hand side is a product of two perfect cubes, because the Hessian is a quadratic polynomial. The identity (2.168) then shows that each factor of the right hand side is proportional to a factor of the expression from the left hand side, and this proportionality can be taken in two ways at will. However, for a fixed choice between those two ways, the proportionality factors should be reciprocal to one another. Indeed, the Hessian can be factorized in infinite many ways as

$$H = mU \cdot \frac{1}{m}V,$$

where U and V are first degree binomials and 'm' is any nonzero number. Thus the identity (2.168) can be written as the system

$$\sqrt{\Delta}f + T = 2m^3U^3, \;\; \sqrt{\Delta}f - T = 2m^{-3}V^3. \tag{2.169}$$

Adding these expressions, one obtains the result (2.160) only in a slightly different form, showing clearly where the external arbitrariness comes into play:

$$\sqrt{\Delta}f = m^3U^3 + m^{-3}V^3. \tag{2.170}$$

One can further decompose the right hand side here into linear factors, to the effect that (2.170) becomes

$$\sqrt{\Delta}f = (mU + m^{-1}V) \cdot (\theta mU + \theta^2 m^{-1}V) \cdot (\theta^2 mU + \theta m^{-1}V).$$

This form allows us to find the roots of the cubic equation $f = 0$ in the form given by equation (2.160) with $k \equiv m^{-2}$. If the roots are all real, k must be complex unimodular as before. For the sake of completeness, we mention that the Jacobian of a cubic can be also obtained from (2.169) as a difference of cubes:

$$T = m^3U^3 - m^{-3}V^3 = (mU - m^{-1}V)(\theta mU - \theta^2 m^{-1}V)(\theta^2 mU - \theta m^{-1}V).$$

This shows that the roots of Jacobian are of the same algebraical nature as the roots of the cubic itself. In formula (2.161) we do not have to change but the sign in both the denominator and numerator in order to get the roots of the corresponding Jacobian. This discussion also shows that the form (2.161) of the roots of a cubic equation is valid independently of the nature of the roots.

It is now important to give a physical interpretation for the external factor k occurring when one wants to construct the cubic given its Hessian. For this we will consider the case where the cubic has real roots, i.e. k is complex of unit modulus. The equations (2.161) for the roots can be written as

$$x_1 = \frac{h + \overline{h}k}{1 + k}, \; x_2 = \frac{h + \varepsilon\overline{h}k}{1 + \varepsilon k}, \; x_3 = \frac{h + \varepsilon^2\overline{h}k}{1 + \varepsilon^2 k} \tag{2.171}$$

with $h, \overline{h}$ the roots of Hessian and $\varepsilon \equiv (-1 + i\sqrt{3})/2$ the cubic root of unity ($i \equiv \sqrt{(-1)}$). Now consider the vector of components x_1, x_2, x_3. This is a 'vector' indeed, but with respect to a special group to be mentioned later. For now, it just happens to represent a real space situation when the three roots are the principal values of a symmetric matrix. We are certainly correct in using this image, at least in a limited way, for there is a space reference frame we can construct in every point of space where the symmetric matrix is physically defined. This is given by three special orthogonal vectors – the principal directions of the symmetric matrix in question. Thus the principal values of such a matrix can be arranged in the column matrix

$$|x\rangle \equiv \begin{pmatrix} x_1 \\ x_2 \\ x_3 \end{pmatrix} \tag{2.172}$$

which can be taken as a vector in matrix representation. Indeed each principal value of the matrix can be interpreted as the component of the vector along the corresponding principal direction.

We can decompose the vector from (2.172) with respect to a plane cutting the axes of our reference frame in the points situated at unit distance from origin. In engineering applications such a plane is called octahedral plane, for it represents one of the faces of an octahedron in space. Assuming therefore the situation in the first octant of our reference frame, the normal component of the vector (2.172) on this plane is, with an obvious notation for transposed vectors, given by

$$|x_n\rangle \equiv |n\rangle\langle n|x\rangle = \frac{1}{\sqrt{3}}\begin{pmatrix}1\\1\\1\end{pmatrix}\frac{1}{\sqrt{3}}(1,1,1)\begin{pmatrix}x_1\\x_2 x_3.\end{pmatrix}$$

The in-plane (tangential) component of (2.172) is then given by

$$|x_1\rangle \equiv |x\rangle - |x_n\rangle = \frac{1}{3}\begin{pmatrix}2x_1 - x_2 - x_3\\2x_2 - x_3 - x_1\\2x_3 - x_1 - x_2\end{pmatrix}$$

It is this last vector, usually called octahedral shear vector in engineering applications – its components are given by the eigenvalues of the so-called deviator of the original matrix – which allows us to interpret the complex number k externally introduced. Namely, the form (2.160) of our cubic allows us to identify its binomial coefficients in terms of the quantities $h, \overline{h}$ and k, up to an arbitrary factor, as

$$\begin{aligned} a_0 &= 1 + k^3,\ \ a_1 = -(h + \overline{h}k^3), \\ a_2 &= h^2 + h^{-2}k^3,\ \ a_3 = -(h^3 + h^{-3}k^3). \end{aligned} \tag{2.173}$$

From this we have right away

$$\frac{1}{3}\sum x_1 = \frac{h + \overline{h}k^3}{1 + k^3} \therefore |x_t\rangle = \frac{(h - \overline{h})k}{1 + k^3}\begin{pmatrix}k - 1\\ \omega(\omega k - 1)\\ \omega^2(\omega^2 k - 1)\end{pmatrix}. \tag{2.174}$$

Now take as reference in the octahedral plane the vector corresponding to $k = 1$, when the roots of the cubic are exclusively determined by the roots of its Hessian, therefore with no arbitrariness whatsoever:

$$|x_t^0\rangle = \frac{(h - \overline{h})}{2}\begin{pmatrix}0\\ \omega(\omega - 1)\\ o^2(\omega^2 - 1)\end{pmatrix}. \tag{2.175}$$

Then calculate the angle of the generic tangential vector with respect to this one by the well-known formula:

$$\cos\phi \equiv \frac{\langle x_t^0|x_t\rangle}{\sqrt{\langle x_t^0|x_t^0\rangle\langle x_t|x_t\rangle}}. \tag{2.176}$$

Using equations (2.174) and (2.175) we get

$$\langle x_t^0 | x_t \rangle = -\frac{3(h-\overline{h})^2 k}{2(1+k^3)}(k+1),$$
$$\langle x_t | x_t \rangle = -6k\frac{(h-\overline{h})^2 k^2}{(1+k^3)^2}, \quad \langle x_t^0 | x_t^0 \rangle = -6\frac{(h-\overline{h})^2}{2^2}$$

so that equation (2.176) becomes

$$\cos\phi \equiv \frac{1}{2}(\sqrt{k} + \frac{1}{\sqrt{k}}). \tag{2.177}$$

We conclude that, indeed, knowing the Hessian does not determine the cubic uniquely. In the case of a cubic with real roots the Hessian actually determines a family of cubics whose roots are defined as a one parameter family. The parameter of this family is given by the angle of orientation of the corresponding octahedral shear vector in the octahedral plane.

The advantage of this last approach to characterizing the cubic equation is mostly physical: as mentioned before, more often than not in physics and engineering problems we have to do with quantities qualifying as coefficients of the Hessian of a cubic function. In this case one can rightfully say that the angle ϕ represents, properly speaking, a gauge freedom. Until a proper geometrization of this statement, let us see however how the Hessian can be built, by using again an example from the deformation theory (for details see the references from [14]). As the Manton's geometrization of Skyrme theory allows us to infer, this construction should be proper in the realm of nuclear matter (for details see [14]). On the other hand, it is perhaps worth mentioning that the deformation of matter, in general, can be properly represented, from a physical point of view, as a gauge process.

The statement that we can measure a certain multidimensional physical magnitude in a certain point of space envisions a highly idealized situation. First of all we are not always able to simultaneously measure multiple physical quantities, in view of the fact that these may interact in such a way that their measurements are mutually exclusive. A well known example is the one of the conjugated variables in quantum mechanics. Nevertheless, we shouldn't go that far with the imagination, for the most obvious example is in the very strain of a continuum. Indeed, while from experimental point of view, we can afford adequate pieces of matter to represent ideal states of strain as close as possible to the standards we desire, inside a continuum the situation changes drastically. One cannot state that in a certain point of that continuum there is a precise state of strain of a kind or another. The most we can think of is a mixture of such states, and even that is a highly idealized situation, for we don't know how the states of strain coexist with each other. But, in the cases where the strain is thought in terms of 3×3 matrices, the reason can always be conducted along the lines that follow (indicated by Novozhilov, for details see the references from [14]).

A matrix quantity defined in a point in space cannot be measured but by its intensities along directions and in planes through that point. The values of these intensities obviously vary with the direction and plan of measurement. However, in a continuum, one can assume that, at least in certain conditions of isotropy, the local manifestation of a matrix quantity is a certain average over all of the possible directions and planes through a point. When the matrix is a symmetric tensor, as one currently assumes in the theory of strains, and furthermore, when one admits a uniform distribution of all directions and planes in space, the averages over directions and planes can be given quite easily.

If $\boldsymbol{x}$ is our matrix, having the eigenvalues $x_{1,2,3}$, then the intensity along a certain direction given by the unit vector $\widehat{n}$, can be calculated with the formula

$$x_n \equiv x_{ij}n^i n^j = x_1(n^1)^2 + x_2(n^2)^2 + x_3(n^3)^2, \tag{2.178}$$

where $n^{1,2,3}$ are the components of the unit vector of direction in the proper system of eigendirections of x. Then we can figure out that, in each one of the space points, a continuum can be characterized by an average of this quantity over the unit sphere. Representing the components of $\widehat{n}$ in terms of spherical angles as usual: $\sin\theta\cos\varphi, \sin\theta\sin\varphi, \cos\theta$, one can assume therefore that the continuum exhibits in any point the mean

$$\overline{x}_n = \frac{1}{4\pi}\oiint_{\text{Unit Sphere}} x_n \sin\theta d\theta d\varphi.$$

Performing this operation in (2.178), one obtains known value

$$\overline{x}_n = \frac{x_1 + x_2 + x_3}{3}.$$

For the cubic (2.36) the previous result takes the value

$$\overline{z} = n/\overline{r} = 1/3. \qquad (2.179)$$

On the other hand, if $\widehat{n}$ is the normal to a plane through a point inside a continuum, we can calculate the intensity of x on this plane, according to the formula representing Pythagoras' theorem

$$x_t^2 \equiv (x^2)_{ij}n^i n^j - (x_{ij}n^i n^j)^2. \qquad (2.180)$$

Using the same procedure of averaging, we can find the average of this quantity in a point of the continuum as:

$$\overline{x_t^2} = \frac{1}{15}\{(x_2 - x_3)^2 + (x_3 - x_1)^2 + (x_1 - x_2)^2\}. \qquad (2.181)$$

For the cubic (2.36) the previous result takes the value

$$\overline{z^2} = \frac{n^2}{\overline{r^2}} = \frac{2}{15}[1 - 3(\frac{nm_0c}{M})^2].$$

It is therefore to be expected that, in a continuum without inhomogeneities, when it comes to the measurement of a tensor, we only have at our disposal the quantities (2.179) and (2.181), in any of its points. And from these two quantities we ought to construct the eigenvalues of the tensor. Obviously then, the tensor is not uniquely defined. Even if the eigenvalues would be at our disposal, we still would have at least the arbitrariness of space rotations in the definition of a tensor. However they are not at our disposal, and we ought to construct them first, using just the quantities (2.179) and (2.180).

In order to do this, we use the previous phase freedom, whereby the case of null phase is well determined by these two quantities. Specifically, in that case, which we take as a reference case, we have for the roots of Hessian

$$u = \overline{x}_n; v = \sqrt{\frac{5}{6}\overline{x_t^2}}; \quad h \equiv u + iv. \qquad (2.182)$$

For the cubic (2.36) the previous results implies:

$$u = 1/3, \ v = \frac{1}{3}\sqrt{1 - 3(\frac{nm_0c}{M})^2}, \ \ h = \frac{1}{3}\left[1 + i\sqrt{1 - 3\left(\frac{nm_0c}{M}\right)^2}\right].$$

One can see that the imaginary part of h is simply proportional with the magnitude of the shear vector in the octahedral plane.

Therefore the orientation of this shear vector in octahedral plane is arbitrary, and this is our gauge freedom. It is eightfold degenerated for there are eight planes in a octahedron. With these results, the formulas from equation (2.171) define a one parameter family of cubics, corresponding to the rotations of the shear vector in its octahedral plane.

However, having now concentrated on the pure mathematical side of the problem, we ought to consider one further algebraic advantage: the values of variables $h, \overline{h}$ and k can be 'scanned' by a simply transitive continuous group with real parameters. Therefore the gauge freedom is way richer than the arbitrary phase lets to be seen. This group has been revealed for the first time by Dan Barbilian (for details see the references from [14]) with the occasion of a study of the Riemannian space associated with the previous family of cubics. We will briefly review Barbilian's theory insisting on some particular technical points necessary for our reference. The basis of approach is the fact that the simply transitive group with real parameters

$$x_k \leftrightarrow \frac{ax_k + b}{cx_k + d}, \quad a, b, c, d \in R,$$

where x_k are the cubic roots previously discussed, induces a simply transitive group for the quantities $h, \overline{h}$ and k, whose action is:

$$h \leftrightarrow \frac{ah + b}{ch + d}, \quad \overline{h} \leftrightarrow \frac{a\overline{h} + b}{ch^* + d}, \quad k \leftrightarrow \frac{c\overline{h} + d}{ch + d}k$$

which will be called from now on Barbilian group. The structure of this group is typical of a $SL(2, R)$ one, which we take in the standard form

$$[B_1, B_2] = B_1, \quad [B_2, B_3] = B_3, \quad [B_3, B_1] = -2B_2, \tag{2.183}$$

where B_k are the infinitesimal generators of the group. Because the group is simply transitive these generators can be easily found as the components of the Cartan frame ([14]) from the formula

$$\begin{aligned} d(f) &\equiv \sum \frac{\partial f}{\partial x^k} dx^k \\ &= \left[\omega^1 \left(h^2 \frac{\partial}{\partial h} + \overline{h}^2 \frac{\partial}{\partial \overline{h}} + (h - \overline{h})k \frac{\partial}{\partial k}\right) + 2\omega^2 \left(h \frac{\partial}{\partial h}\right.\right. \\ &\quad \left.\left. + \overline{h} \frac{\partial}{\partial \overline{h}}\right) + \omega^3 \left(\frac{\partial}{\partial h} + \frac{\partial}{\partial \overline{h}}\right)\right](f), \end{aligned} \tag{2.184}$$

where ω^k are the components of the Cartan coframe to be found from the system

$$\begin{aligned} dh &= \omega^1 h^2 + 2\omega^2 h + \omega^3, \\ d\overline{h} &= \omega^1 \overline{h}^2 + 2\omega^2 \overline{h} + \omega^3, \\ dk &= \omega^1 k(h - \overline{h}). \end{aligned}$$

Thus we have immediately both the infinitesimal generators and the coframe by identifying the right hand side of equation (2.184) with the standard dot product of $SL(2, R)$ algebra:

$$\omega^1 B_3 + \omega^3 B_1 - 2\omega^2 B_2$$

so that

$$\begin{aligned} B_1 &= \frac{\partial}{\partial h} + \frac{\partial}{\partial \overline{h}}, \quad B_2 = h\frac{\partial}{\partial h} + \overline{h}\frac{\partial}{\partial \overline{h}}, \\ B_3 &= h^2\frac{\partial}{\partial h} + \overline{h}^2\frac{\partial}{\partial \overline{h}} + (h - \overline{h})k\frac{\partial}{\partial k} \end{aligned} \tag{2.185}$$

and

$$\begin{aligned} \omega' &= \frac{dk}{(h - \overline{h})k}, \quad 2\omega^2 = \frac{dh - \overline{h}}{h - \overline{h}} - \frac{h + \overline{h}}{h - \overline{h}}\frac{dk}{k} \\ \omega^3 &= \frac{hd\overline{h} - \overline{h}dh}{h - \overline{h}} + \frac{h\overline{h}}{h - \overline{h}}\frac{dk}{k}. \end{aligned}$$

Since $h = u + iv$, $k = e^{i\phi}$, these last equations can be written as

$$\begin{aligned} B_1 &= \frac{\partial}{\partial u}, B_2 = u\frac{\partial}{\partial u} + v\frac{\partial}{\partial v}, \ B_3 = (u^2 - v^2)\frac{\partial}{\partial u} + 2uv\frac{\partial}{\partial v} + 2v\frac{\partial}{\partial \phi}, \\ \omega' &= \frac{d\phi}{2v}, \ \omega^2 = \frac{dv}{v} - \frac{u}{v}d\phi, \ \omega^3 = \frac{u^2 + v^2}{2v}d\phi + \frac{vdu - udv}{v}. \end{aligned} \tag{2.186}$$

It should be mentioned that, in his original paper, Barbilian does not work with the above differential forms but with the absolute invariant differentials

$$\omega^1 = \frac{dh}{(h - \overline{h})k}, \quad \omega^2 = -i\left(\frac{dk}{k} - \frac{dh + d\overline{h}}{h - \overline{h}}\right), \quad \omega^3 = -\frac{kd\overline{h}}{h - \overline{h}}, \tag{2.187}$$

or, in real terms, exhibiting a three-dimensional Lorentz structure of this space

$$\Omega^1 \equiv \omega^2 = d\phi + \frac{du}{v},\ \Omega^2 = \cos\phi\frac{du}{v} + \sin\phi\frac{dv}{v},$$
$$\Omega^3 = -\sin\phi\frac{du}{v} + \cos\phi\frac{dv}{v}. \tag{2.188}$$

The advantage of this representation is that it makes obvious the connection with the Poincaré representation of the Lobachevsky plane. Indeed, the metric here is

$$-(\Omega^1)^2 + (\Omega^2)^2 + (\Omega^3)^2 = -(d\phi + \frac{du}{v})^2 + \frac{(du)^2 + (dv)^2}{v^2}. \tag{2.189}$$

This metric reduces to that of Poincaré in case $\Omega^1 = 0$ which, as Barbilian noticed, defines the variable ϕ as the 'angle of parallelism' of the hyperbolic plane (the connection). In fact, recalling that in modern terms (du/v) represents the connection form of the hyperbolic plane, the equations (2.188) then represent a general Bäcklund transformation in that plane [14].

In view of the importance that we revealed for the geometry of Lobachevsky in the classical problems of motion, it becomes also important to know the meaning of the condition $\Omega^1 = 0$ for a family of cubic equations [14]. It turns out that it expresses the so-called apolar transport of cubics. This transport is defined by the condition that any root of the transported cubic is in a harmonic relation with any root of the original cubic, with respect to the other two remaining roots of this last one:

$$\frac{y_l - x_j}{y_l - x_k} : \frac{x_i - x_j}{x_i - x_k} = -1, \quad i \neq j \neq k \neq i \tag{2.190}$$

in all positive permutations of the indices i, j, k and for every l. Therefore each new root (y_l) and each of the corresponding old ones (x_i), are in harmonic range with respect to the other two old roots (x_j, x_k). Then it can be proved that the conditions from equation (2.190) boil down to the vanishing of the bilinear invariant of the two cubics, analogous to the bilinear invariant of the quadratics:

$$a_3b_0 - 3a_2b_1 + 3a_1b_2 - a_0b_3. \tag{2.191}$$

Here a_m denote the coefficients of the original cubic, while b_m denote the coefficients of the transported cubic. Obviously, this invariant is analogous to the one from the case of two quadratics, whose vanishing expresses the fact that their roots are in harmonic sequence. The geometry related to this invariant is century old (for details see references from [14]) and Dan Barbilian seemed particularly fond of it (for details see references from [14]), for he elaborated for a long while on its different aspects, especially related to the geometry of the triangle. As the triangle comes nowadays in relation with the construction of skyrmions from instantons (for details see references from [14]), from a point of view closely related to its geometry, it is therefore worth considering this connection, which turns out to be strictly related to the physics of continua.

Now, if the two cubics are infinitesimally close, then the condition of their transport by involution reduces to

$$a_3 da_0 - 3a_2 da_1 + 3a_1 da_2 - a_0 da_3 = 0. \tag{2.192}$$

For the cubic (2.36) the apolar transport implies

$$E = \pm m_0 c^2 \sqrt{\frac{2}{3} - \text{const.} \left(\frac{M}{n m_0 c}\right)^2}.$$

Using here the equations (2.173) above for the coefficients, the condition of apolar transport of the cubics amounts to

$$-(h - \bar{h})^3 k^3 \left(\frac{dk}{k} - \frac{dh + d\bar{h}}{h - \bar{h}}\right) = 0.$$

As the cubics are asumed to have distinct roots, this condition is satisfied if, and only if, the differential form Ω^1 is null. Therefore the parallel transport of the hyperbolic plane actually represents the apolar transport of the cubics.

This way, the vectors represented by the real eigenvalues of a certain matrix, have a sure physical interpretation in the framework of the classical theory of motion. Assume, for the sake of exemplification a hydrogen atom: it is described by a single motion. We have seen that the nuclear matter in such representation can be characterized by a complex number depending on the eccentricity and the orientation of the orbit ([14]). This complex number can be assumed to represent a particular state of stress inside nucleus, given by equations (2.182) above. The eigenvalues of stress are then given by Barbilian formulas using the gauge freedom. However, insofar as they are supposed to be measured values themselves, they reveal an outstanding meaning of the root of Hessian: it is the (complex) parameter of a Cauchy distribution.

Peter McCullagh has noticed a curious property of the one-dimensional Cauchy distribution, which is related to the benefit of a complex parameterization of this distribution (for details see the references from [14]). The parameters of a statistical distribution are usually taken as real, but McCullagh shows a clear advantage of representing them in a complex form, at least when it comes to Cauchy distribution. He starts with the fact that this distribution for a single variate X can be written in the form

$$f_x(x|\theta) = \frac{|\theta_2|}{\pi|x-\theta|^2}; \quad \theta \equiv \theta_1 + i\theta_2, \tag{2.193}$$

where θ is the 'complex parameter' of the distribution. The real part of this parameter gives the location of data, while the imaginary part roughly characterizes the spread of the distribution. One knows that this class of distributions is closed with respect to the homographic transformation of the variable: any linear fractional transform of X has also a Cauchy distribution. But the complex representation of the parameter brings to light one of the most important consequences of this theorem: if X belongs to the Cauchy class with the complex parameter θ, i.e. symbolically $X \sim C(\theta)$, then we have

$$\frac{aX+b}{cX+d} \sim C\left(\frac{a\theta+b}{c\theta+d}\right). \tag{2.194}$$

This property allows us to give efficient estimators for the complex parameter θ, based on the principle of maximum likelihood.

As a rule, the likelihood function used in estimations is simply the product of the values of the probability density for the different measured values of X. In taking the maximum likelihood with respect to parameters, it would be therefore appropriate to work with the logarithm of the likelihood, and this is what practically happens. For instance if one measures two values of X having the probability density (2.193), say x_1 and x_2, the likelihood function constructed based on this information is simply:

$$L(\theta|x_1,x_2) = \frac{\theta_2^2}{\pi^2|x_1-\theta|^2|x_2-\theta|^2}. \tag{2.195}$$

The likelihood is maximum with respect to θ when the derivatives of this function with respect to θ_1 and θ_2 are null. In terms of the log-likelihood, which is a lot easier to handle, we then have:

$$\frac{\partial}{\partial\theta_1}\ln L(\theta|x_1,x_2) = \frac{\partial}{\partial\theta_2}\ln L(\theta|x_1,x_2) = 0. \tag{2.196}$$

In view of the fact that

$$\ln L(\theta|x_1,x_2) = -2\ln\pi + 2\ln|\theta_2| - \sum_i \ln(x_i-\theta) - \sum_i \ln(x_i-\overline{\theta}), \tag{2.197}$$

where the sumation extends over the two measured values and a star denotes complex conjugation, the two equations (2.196) become:

$$\sum_i \frac{1}{x_i-\theta} + \sum_i \frac{1}{x_i-\overline{\theta}} = 0, \quad \frac{2}{i\theta_2} + \sum_i \frac{1}{x_i-\theta} - \sum_i \frac{1}{x_i-\overline{\theta}} = 0. \tag{2.198}$$

Therefore the sum here is a purely imaginary number, as we assume that the values xi are real. The second one of these equation shows that

$$\sum_i \frac{1}{x_i-\theta} = -\frac{1}{i\theta_2}. \tag{2.199}$$

If we sum up here and clear the denominators, we get

$$i\theta_2(x_1+x_2-2\theta_1-2i\theta_2) = -x_1x_2+(x_1+x_2)(\theta_1+i\theta_2)-\theta_1^2+\theta_2^2-2i\theta_1\theta_2. \tag{2.200}$$

Solving this equation shows what one already knows well about the Cauchy distribution. First, with the information of only two measured values we cannot have an estimation for the mean; it can be any value between the two measured ones. As to the variance estimator, it is also indeterminate, but this is quite a natural characteristic, so to speak, of this type of repartition, because it has no finite moments of higher order.

At this point we can easily see the advantage of equation (2.194): it shows that the best determination of the Cauchy distribution involves just as many measured values of X, as the determination of a real linear-fractional or Möbius, in terms of McCullagh, transformation. Therefore we need to have three measurements of the statistical variable X, in order to determine a Cauchy distribution the best possible way. The general estimator will then be calculated from a particularly convenient Cauchy distribution through a well-defined transformation. Let us do some calculations.

In equations (2.198) and (2.199) nothing changes, except the fact that the sum should be now performed on three values of X, say x_1, x_2, x_3, instead of two. So, instead of (2.198) we have

$$\sum_i \frac{1}{x_i - \theta} + \sum_i \frac{1}{x_i - \overline{\theta}} = 0, \quad \frac{3}{i\theta_2} + \sum_i \frac{1}{x_i - \theta} - \sum_i \frac{1}{x_i - \overline{\theta}} = 0 \tag{2.201}$$

and instead of (2.199) we have

$$\sum_i \frac{1}{x_i - \theta} = -\frac{3}{2i\theta_2}, \tag{2.202}$$

as well as the complex conjugate of this equation. Now, the direct calculation of the estimators for θ_1 and θ_2 is rather tedious. Nevertheless, we can simplify it, using the property (2.194). The procedure among to choosing three particular values for X, say $-1, 0, 1$, and calculate the estimator of θ for them; then take the homographic transform of this estimator through the homography that carries $-1, 0, 1$, into the values x_1, x_2, x_3 of X. Indeed such a real homography is well determined. Let us consider that the values (x_1, x_2, x_3) do correspond to the values $(-1, 0, 1)$ in this order. If the matrix of this homography has the entries a, b, c, d, then we can find it up to a normalization factor from the system of equations

$$x_1 = \frac{-a+b}{-c+d}, \quad x_2 = \frac{b}{d}, \quad x_3 = \frac{a+b}{c+d}. \tag{2.203}$$

This gives

$$\frac{a}{x_2x_3 + x_1x_2 - 2x_1x_3} = \frac{b}{x_2(x_3 - x_1)} = \frac{c}{2x_2 - x_3 - x_1} = \frac{d}{x_3 - x_1}. \tag{2.204}$$

The problem is now to find the estimator θ for the particular values $(-1, 0, 1)$. This can be easily done from equation (2.202) and its complex conjugate, which give the system

$$\theta_1 = 0, \quad 3\theta_2^2 = 1. \tag{2.205}$$

Therefore, in this particular case we have simply $i/\sqrt{(3)}$ as an estimation for the parameter θ: it is purely imaginary. The estimator according to arbitrary data (x_1, x_2, x_3) will then be obtained through the homography given by equations (2.204):

$$\theta = \frac{(x_2x_3 + x_1x_2 - 2x_1x_3)\frac{i}{\sqrt{3}} + x_2(x_3 - x_1)}{(2x_2 - x_3 - x_1)\frac{i}{\sqrt{3}} + (x_3 - x_1)}. \tag{2.206}$$

In real terms we have:

$$\theta_1 = \frac{\sum x_1(x_2 - x_3)^2}{\sum(x_2 - x_3)^2}, \quad \theta_2 = \sqrt{3}\frac{(x_2 - x_3)(x_3 - x_1)(x_1 - x_2)}{\sum(x_2 - x_3)^2}. \tag{2.207}$$

Therefore, the complex estimator of the Cauchy distribution is in close relationship with the Hessian of the cubic having the roots (x_1, x_2, x_3). More to the point it is the root of that Hessian. Indeed, in terms of the roots of a cubic equation its Hessian is:

$$\{\sum(x_2 - x_3)^2\}x^2 - 2\{\sum x_1(x_2 - x_3)^2\}x + \{\sum x_1^2(x_2 - x_3)^2\} = 0 \tag{2.208}$$

and its roots are θ above and its complex conjugate. The expression from equation (2.207) are the real and imaginary parts of these roots. Even more, the sum and the product of the two complex estimators are given by the mean and the standard deviation of the three values, with respect to the system of probabilities:

$$p_1 \equiv \frac{(x_2 - x_3)}{\sum(x_2 - x_3)^2}, \quad p_2 \equiv \frac{(x_3 - x_1)^2}{\sum(x_2 - x_3)^2}, \quad p_3 \equiv \frac{(x_1 - x_2)^2}{\sum(x_2 - x_3)^2}, \tag{2.209}$$

which they determine quite naturally.

Therefore, in this statistical interpretation, the root of the Hessian is the parameter of a Cauchy distribution. The roots of the corresponding cubic are three measurements of the Cauchy variate, that give the most reliable estimation of the parameter. The problem with this representation of the connection between cubic and its Hessian is that the Cauchy distribution is referring to a one-dimensional variate. However, everything gets in order if we take the Cauchy density of probability as a marginal distribution of a Gaussian in plane (for details see [14]).

References

1. Adler, R., Bazin, M., Schiffer, M.: Introduction to General Relativity, Mc Graw Hill, (1972).
2. Agop, M.: Acta Physica Polonica, A 64, 251, (1982).
3. Agop, M.: The use of elliptic functions in the study of some interactions, especially electric and gravitational (PhD Thesis, in Romanian), Iasi, (1983) (not published).

4. Agop, M., Mehedinţeanu, S.: Revue Roumaine de Physique, (1986), 31, 3, 201.

5. Agop, M., Mazilu, N.: Revue Roumaine de Physique, 31, 4, 305, (1986).

6. Appel, P., Lacour, E.: Principles de la théorie des fonctions et applications, Gauthier Villars, Paris, 2-éme (1926).

7. Einstein, A.: Relativity: The Special and the General Theory, Barnes ans Noble Library of Essential Reading, (2004).

8. Fock, V.A.: The Theory of Space, Time and Gravitation, Macmillan, (1964).

9. Ionescu-Pallas, N.: General Relativity and Cosmology, Scientific and Encyclopedic Publishing House, Bucharest, (1980).

10. Ionescu-Pallas, N., Sofonea, L.: Revue Roumaine de Mathematique Pure and Applique, 27, 1, 33, 1982.

11. Kramer, D. et al. (E Schmutzer, Editor), Exact solutions of Einstein's field equations, Cambridge University Press, (1980).

12. Kruskal, M.D.: Physical Review, 119, 1743, (1960).

13. Landau, L.D., Lifshitz, E.M.: Théorie des champs, Mir, Moscou, (1970).

14. Mazilu, M., Agop, M.: Skyrmions. A great Finishing Touch to Classical Newtonian Philosophy, Nova Publishers, New York, (2012).

15. Milne, E.A.: Kinematic Relativity, Oxford Clarendon Press, (1948).

16. Onicescu, O.: Invariantive Mechanics, Springer Verlag, Wien, (1975).

17. Sachs, E.: Physical Review, 128, 2851, (1962).

18. Yamamoto, T.: Progress in Theoretical Physics, Japan, 8, 258, (1952).

CHAPTER 3

The gravitational and stationary electro-magnetic fields

3.1 The complex potential

In the previous chapter Einstein's outlook on geodesic motion for various metrics with spherical symmetry, has been presented. In our opinion, investigation by means of elliptic functions and integrals has the gift of being most general, evidentiating particularities of each metric (in fact, its source) through the medium of degeneracy limits of elliptic functions.

The purpose of this chapter is to explicitate another Einsteinian principle, leading to the metric definition itself, for details see [1]-[17]. Namely, the metric tensor g_{ik} is determined by the energy tensor of the space by means of Einstein's field equations which, after a convenient adjustment of the Einsteinian gravitational constant and metric signature (more precisely, in an appropriately chosen system of physical units) are written as:

$$G_{ik} = -6\pi T_{ik}, \quad i,k = 0,1,2,3. \tag{3.1}$$

In the relation (3.1) T_{ik} is the energy tensor and G_{ik} Einstein's tensor, written in terms of the contracted curvature tensor R_{ik} and curvature invariant R as:

$$G_{ik} = R_{ik} - \frac{1}{2}Rg_{ik}. \tag{3.2}$$

Postulation of equations (3.1) in order to determine the metric tensor has, as a consequence, the fact that their solutions are non-unique, not even in the most simple cases. That is why one usually speaks about "classes of solutions" of Einstein's equations and, correspondingly, about "classes of metrics". These classes are characterized by aprioric conditions: flatness at infinity, symmetries, etc (for details, see [18]-[32]). Even in the most simple cases, such as vacuum or stationary electromagnetic vacuum, there is a multitude of solutions, with various known metrics as particular cases. Nevertheless, in many cases appears possibility of their rational "classification", meaning that they can be obtained one from the other by means of certain groups of transformation.

To put them into evidence, we shall use the method of "complex potentials", or "Ernst potentials", whose mechanism is particularly well described by W. Israel and G.A. Wilson [18]. In this respect, let us write the space-time metric in the form:

$$ds^2 = f(dx^0 + \omega_\alpha dx^\alpha)^2 - f^{-1}\gamma_{\alpha\beta} dx^\alpha dx^\beta, \tag{3.3}$$

where summation over repeated indices within a term is understood, and we denoted:

$$\begin{aligned} \gamma_{\alpha\beta} &= g_{0\alpha} g_{0\beta} - g_{00} g_{\alpha\beta}, \\ \omega_\alpha &= g_{0\alpha}/g_{00}, \\ f &= g_{00}. \end{aligned} \tag{3.4}$$

Here $(\gamma_{\alpha\beta})$ is a three-dimensional metric which can be used for raising and lowering indices. The three-vector ω_α is defined only up to a gradient, corresponding to the arbitrary choice of the origin of the temporal coordinate x^0. So, it can be used to obtain a vector invariant with respect to the arbitrary choice of the origin, by means of operation "curl" in the metric $\gamma_{\alpha\beta}$, which is:

$$\frac{1}{f^2}\boldsymbol{\tau} = -\text{curl}\boldsymbol{\omega}, \tag{3.5}$$

or, explicitly:

$$\frac{1}{f^2}\tau^\alpha = -\frac{1}{\sqrt{\gamma}}\varepsilon^{\alpha\beta\mu}\partial_\beta\omega_\mu, \tag{3.6}$$

where $\gamma = \det(\gamma_{\alpha\beta})$, and $\varepsilon_{\alpha\beta\mu}$ is the three-dimensional totally anti-symmetric symbol.

Let us now suppose that in the space-time of metric (3.3) exists an electromagnetic stationary field, deriving from the potential A_k:

$$F_{ik} = \partial_k A_i - \partial_i A_k. \tag{3.7}$$

By virtue of time independence, the electric components of the electromagnetic field tensor are:

$$F_{0\alpha} = \partial_\alpha A_0, \tag{3.8}$$

while Maxwell's equations write:

$$\partial_\alpha\{(-g)^{1/2}F^{i\alpha}\} = 0, \tag{3.9}$$

where $g = \det(g_{ik})$ can be written as:

$$(-g)^{1/2} = f^{-1}\gamma^{1/2}. \tag{3.10}$$

Let us now show that the full set of Maxwell's equations can be reduced to a single equation with partial derivatives for a scalar complex field. To this end, one observes that (3.9), with $i = \beta$, makes it possible to write $F^{\beta\alpha}$ in terms of a "magnetic" potential ϕ as follows:

$$F^{\beta\alpha} = f\gamma^{-1/2}\varepsilon^{\beta\alpha\mu}\partial_\mu\phi. \tag{3.11}$$

The rest of components can then be expressed by means of (3.8) and (3.11), according to:

$$F^{\alpha 0} = \omega_\beta F^{\beta\alpha} + F_{0\beta}\gamma^{\beta\alpha}. \tag{3.12}$$

In view of (3.11), (3.8), and (3.6), equation (3.9) for $i = 0$ becomes:

$$\operatorname{div}(f^{-1}\nabla A_0) = -f^{-2}\boldsymbol{\tau}\cdot\nabla\phi. \tag{3.13}$$

Using cyclic identity:

$$\varepsilon^{\alpha\beta\mu}\partial_\mu F_{\alpha\beta} = 0,$$

where $F_{\alpha\beta}$ is given by (3.11), we still have:

$$\operatorname{div}(f^{-1}\nabla\phi) = f^{-2}\boldsymbol{\tau}\cdot\nabla A_0. \tag{3.14}$$

We mention, again, that all the differential operators are expressed in terms of the three-dimensional metric $\gamma_{\alpha\beta}$.

Next, let us define the complex potential:

$$\psi = A_0 + i\phi\ , \quad i = \sqrt{-1}. \tag{3.15}$$

By means of (3.13)–(3.15), we then have:

$$\operatorname{div}(f^{-1}\nabla\psi) = f^{-2}\boldsymbol{\tau}\cdot\nabla\psi, \tag{3.16}$$

expressing the whole set of Maxwell's equations.

In an analogous way, one can express the Ricci tensor R_{ik} by means of the vector $\boldsymbol{G}$ defined as:

$$2f\boldsymbol{G} = \nabla f + i\boldsymbol{\tau}. \tag{3.17}$$

The components of R_{ik} are then defined by:

$$-f^{-2}R_{00} = \mathrm{div}\boldsymbol{G} + \boldsymbol{G}\cdot(\boldsymbol{G}^* - \boldsymbol{G}), \tag{3.18}$$

$$-2if^{-2}R_0^{\alpha} = \gamma^{-1/2}\varepsilon^{\alpha\beta\mu}(\partial_\mu G_\beta + G_\beta G_\mu^*) \ , \tag{3.19}$$

$$f^{-2}(\gamma_{\mu\alpha}\gamma_{\nu\beta}R^{\alpha\beta} - \gamma_{\mu\nu}R_{00}) = R_{\mu\nu}(\gamma) + G_\mu G_\nu^* + G_\mu^* G_\nu \ . \tag{3.20}$$

Here $R_{\mu\nu}(\gamma)$ denotes the Ricci tensor of the three-dimensional metric $(\gamma_{\alpha\beta})$, and by asterisk (*) is marked the complex conjugate. The energy tensor of the electromagnetic field is then given by general formula:

$$-4\pi T_{ij} = g^{lm}F_{im}F_{jl} - \frac{1}{4}g_{ij}F_{lm}F^{lm}.$$

Since, according to our previous results:

$$\frac{1}{2}F_{ik}F^{ik} = (\nabla\phi)^2 - (\nabla A_0)^2,$$

we still have:

$$8\pi f^{-1}T_{00} = (\nabla\phi)^2 + (\nabla A_0)^2, \tag{3.21}$$

$$4\pi f^{-1}T_0^{\alpha} = \gamma^{-1/2}\varepsilon^{\alpha\mu\nu}(\partial_\mu\phi)(\partial_\nu A_0), \tag{3.22}$$

$$\begin{aligned} -\,4\pi f^{-1}T^{\alpha\beta} &= (\partial^\alpha\phi)(\partial^\beta\phi) + (\partial^\alpha A_0)(\partial^\beta A_0) \\ &-\frac{1}{2}\gamma^{\alpha\beta}[(\nabla\phi)^2 + (\nabla A_0)^2]. \end{aligned} \tag{3.23}$$

Since the energy tensor of the electromagnetic field is a null trace tensor [1], Einstein's equations (3.1) can be written in a simpler form as:

$$R_{ik} = -8\pi T_{ik}.$$

Combining these equations with (3.19) and (3.22), we have:

$$\mathrm{curl}\boldsymbol{\tau} = -4(\nabla\phi \times \nabla A_0) = i\,\mathrm{curl}(\psi\nabla\psi^* - \psi^*\nabla\psi),$$

so that the vector quantity:

$$\boldsymbol{\tau} + i(\psi^* \nabla \psi - \psi \nabla \psi^*)$$

derives from a potential, say φ, defined up to an arbitrary additive constant:

$$\boldsymbol{\tau} + i(\psi^* \nabla \psi - \psi \nabla \psi^*) = \nabla \varphi. \tag{3.24}$$

Let us now define a new complex potential:

$$E = f - \psi \psi^* + i\,\varphi, \tag{3.25}$$

so that (3.24) and (3.17) yield:

$$f\boldsymbol{G} = -\frac{1}{2}\nabla E + \psi^* \nabla \psi. \tag{3.26}$$

Substituting (3.26) in (3.18) and (3.21), by means of (3.16), we still have:

$$f\nabla^2 E = \nabla E(\nabla E + 2\psi^* \nabla \psi). \tag{3.27}$$

We also observe that (3.16) can be written in a form analogous to (3.27), that is:

$$f\nabla^2 \psi = \nabla \psi(\nabla E + 2\psi^* \nabla \psi). \tag{3.28}$$

By means of the complex potentials E and ψ, and the metric $\gamma_{\alpha\beta}$ as well, one can express everything. According to (3.25), we have:

$$f = \frac{1}{2}(E + E^*) + \psi^* \psi, \tag{3.29}$$

while (3.20) and (3.23) yield:

$$\begin{aligned} - f^2 R_{\alpha\beta}(\gamma) &= \frac{1}{2}(E, E^*)_{\alpha\beta} + \psi(E, \psi^*)_{\alpha\beta} \\ &+ \psi^*(E^*, \psi)_{\alpha\beta} - (E + E^*)(\psi, \psi^*)_{\alpha\beta}, \end{aligned} \tag{3.30}$$

where we denoted:

$$2(A, B)_{\alpha\beta} \equiv (\partial_\alpha A)(\partial_\beta B) + (\partial_\beta A)(\partial_\alpha B).$$

Therefore, the complete system of equations of gravitational and electromagnetic fields in vacuum is now expressed by (3.27), (3.28), and (3.30).

The idea of using a complex potential in order to express the equations of the gravitational field belongs to F.J.Ernst, first time concerning the axially-symmetric metric ([10], [11]), and then for the general case of a stationary metric [12]. This last case of Ernst's investigation concerns the gravitational field in vacuum, being comprised in equations:

$$\frac{1}{2}(\varepsilon+\varepsilon^*)\nabla^2\varepsilon = \nabla\varepsilon\cdot\nabla\varepsilon, \tag{3.31}$$

and:

$$-(\varepsilon+\varepsilon^*)R_{\alpha\beta}(\gamma) = (\partial_\alpha\varepsilon)(\partial_\beta\varepsilon^*) + (\partial_\beta\varepsilon)(\partial_\alpha\varepsilon^*) \tag{3.32}$$

which are obtained from (3.27), (3.28), and (3.30), by setting $\psi = 0$ and $\varepsilon = f + i\varphi$ in (3.25).

The importance of this approach to solution of Einstein's equations is revealed by the fact that the way of obtaining equations (3.27), (3.28) and (3.30) (or (3.31) and (3.32) for the pure gravitational case) can be somehow inverted. Namely, $(\gamma_{\alpha\beta})$ can be any three-dimensional metric, conveniently chosen, for which (3.30) and (3.32) represent the curvature tensor. This way, our attention is focused on solving (3.27) and (3.28), or Ernst's equation (3.31), intensively investigated by several authors (see, *e.g.*, [20]).

The general procedure of construction of the metric (3.3) would be to obtain f, $\boldsymbol{\omega}$ from the solutions of (3.27), (3.28), and (3.30), followed by A_0 and ϕ, and, finally, the components of the electromagnetic field. The most difficult choice proves to be the metric $(\gamma_{\alpha\beta})$: it has to possess a certain liberty, so as to be compatible with equations (3.32) for the purely gravitational field. Otherwise, the use of equations (3.30) and (3.32) is not essential: there is a theorem concerning generation of families of solutions, starting from their existence for the purely gravitational field (see, *e.g.*, [26]).

Next, we are going to put into evidence a certain procedure of solutions generation, starting with already known solutions, as it was shown by Kinnersley [20]. We shall follow his investigation, but restricting ourselves only to the bidimensional case. To this end, let us introduce two complex potentials, u and v in (3.31), instead of complex potential ε, by means of:

$$\varepsilon = \frac{u-v}{u+v}. \tag{3.33}$$

Obviously, one of the potentials u and v is redundant, and their ratio is the only thing that counts. In other words, it is up to our choice to take one of potentials u and v so that definition (3.33) be compatible with Ernst's equation (3.31). Introducing (3.33) into (3.31), we find as sufficient condition the set of symmetric differential equations:

$$(uu^* - vv^*)\nabla^2 u = 2(u^*\nabla u - v^*\nabla v)\nabla u,$$
$$(uu^* - vv^*)\nabla^2 v = 2(u^*\nabla u - v^*\nabla v)\nabla v. \tag{3.34}$$

This definition allows us to introduce a complex bidimensional space, of vector components u and v:

$$z = (u, v)$$

metrized by the indefinite quadratic form with metric matrix:

$$\eta_{\alpha\beta} = \mathrm{diag}(1, -1), \quad \alpha, \beta = 1, 2. \tag{3.35}$$

In this case, equations (3.34) can be written in a condensed form as:

$$z^\alpha z^*_\alpha \nabla^2 z^\beta = 2 z^*_\alpha \nabla z^\alpha \nabla z^\beta. \tag{3.36}$$

This is a vector equation whose group of covariants leave invariant the metric (3.35), that is invariates the Hermitian scalar products. Therefore, we have to do with the unitary matrix group 2×2:

$$\begin{bmatrix} a & b \\ -b^* & a^* \end{bmatrix}, \tag{3.37}$$

subject to condition:

$$aa^* + bb^* = 1.$$

We mention, nevertheless, that by virtue of (3.33) this group acts on Ernst's potential as a homographic group: denoting $\varepsilon = (\xi - 1)/(\xi + 1)$, then the complex homography:

$$\xi' = \frac{a\xi - b^*}{b\xi + a^*} \tag{3.38}$$

shall correspond to transformation given by matrix (3.37). Due to this fact, matrix (3.37) is defined only up to an arbitrary normation factor.

The property of Ernst's equation of being invariant with respect to the homography can also be made obvious in a straight manner, but the way used above puts into evidence an important fact: the Ernst potential can be considered as an inhommogeneous coordinate for a certain binary domain [6]. This fact shall subsequently be explained, by means of some operational significance given to this coordinate.

3.2 A variational principle

The existence of theorems regarding generation of solutions of the field equations with the help of already known solutions [26], allows us to restrict our investigation to Ernst's equation (3.31), as being of maximum generality. Ernst [12] mentions two essential points regarding this equation. First, it is invariant under complex conjugation: the complex conjugate of ε satisfies the same equations and, therefore, can play the role of the complex gravitational potential. Second, Ernst evidentiates the fact that the formal equation (3.33) can be derived from a variational principle. This last aspect is not very much approached in the literature, even if, in our opinion, it is of a maximum formal importance. Indeed, equation (3.33) is one of the Euler-Lagrange equations associated with the functional:

$$2\int \frac{\nabla\varepsilon\nabla\varepsilon^*}{(\varepsilon+\varepsilon^*)^2}\gamma^{1/2}\,d^3x, \tag{3.39}$$

the other one being its complex conjugate. Here we mention, once again, that the dot product is understood in the "ambient" metric $(\gamma_{\alpha\beta})$.

The way of obtaining equation (3.31) did not *simultaneously* reveal its complex conjugate, its validity being expressed as an *a posteriori* observation. But the variational principle attached to functional (3.39) does not make any difference between them.

In view of these observations, appears the problem of validity of variational principle itself, since the existence of the complex conjugate equation exceeds the frame within which it has been deduced. Regarding Einstein's equations, only one of these two variants can be used.

The answer to this problem can be found at the end of the previous paragraph. As we have seen, the equation is invariant with respect to the group of homographies – a continuous group with three essential parameters [6]. Without loss of generality, we shall consider these homographies as being real; this is always possible, and can be performed by applying a Cayley-type [14] transformation to matrix (3.37), in order to make it real. In addition, we shall use a complex variable h, instead of variable ξ, and denote by $\overline{h}$ the complex conjugate of h.

Consider, therefore, a real homographic transformation on the variable h:

$$\begin{aligned} h' &= \frac{ah+b}{ch+d}, \quad a, b, c, d \in \Re, \\ ad - bc &\neq 0. \end{aligned} \tag{3.40}$$

The conjugates of the too complex quantities h and h' are connected, obviously, by the the same transformation:

$$\overline{h'} = \frac{a\overline{h}+b}{c\overline{h}+d}. \tag{3.41}$$

Transformations (3.40) and (3.41) form a group of two variables with three parameters (one of the quantities a, b, c, d is not essential), as it can easily be verified. This group leaves invariant the differential quadratic form:

$$-4\frac{dh\, d\overline{h}}{(h-\overline{h})^2}. \tag{3.42}$$

Indeed, in view of (3.40) and (3.41), we have:

$$dh' = \frac{(ad-bc)dh}{(ch+d)^2},$$

$$d\overline{h'} = \frac{(ad-bc)d\overline{h}}{(c\overline{h}+d)^2},$$

$$h' - \overline{h'} = \frac{(ad-bc)(h-\overline{h})}{(ch+d)(c\overline{h}+d)},$$

which immediately yields:

$$\frac{dh' d\overline{h'}}{(h'-\overline{h'})^2} = \frac{dh\, d\overline{h}}{(h-\overline{h})^2}.$$

The quadratic form (3.42) is a well-known metric, namely metric of the Lobachevsky plane, in Poincaré's representation [9]. It can be written under a real form, which is most commonly encountered, by explicitating $h = u + iv$. The result is:

$$-4\frac{dh\,d\overline{h}}{(h-\overline{h})^2} = \frac{du^2 + dv^2}{v^2}. \tag{3.43}$$

Recently, the properties of this metric have been emphasized in connection with some theories on supergravity [30]. We could be interested in its behavior as far as h would represent an Ernst potential, in which case the group (3.40)–(3.41) would continuously generate solutions. Now, we have to answer the question: is there some equation with all these solutions? Such an equation is necessary to obtain at least one of the concrete forms of h, the other ones being then generated through the medium of (3.40).

Supposing that the "ambient" "metric" is $\gamma_{\alpha\beta}$, the Lagrangian function corresponding to metric (3.42), according to the usual procedure [22], writes:

$$L = \gamma^{\alpha\beta}\frac{\frac{\partial h}{\partial x^\alpha}\frac{\partial \overline{h}}{\partial x^\beta}}{(h-\overline{h})^2} = \frac{\nabla h \nabla \overline{h}}{(h-\overline{h})^2}.$$

The variational principle:

$$\delta \int L\gamma^{1/2}\, d^3x = 0 \tag{3.44}$$

leads then to the equations:

$$\begin{aligned}(h-\overline{h})\nabla^2 h &= 2\nabla h \nabla h,\\ (h-\overline{h})\nabla^2 \overline{h} &= 2\nabla \overline{h} \nabla \overline{h}.\end{aligned} \tag{3.45}$$

For $h = i\varepsilon$, we are left with Ernst's equations.

The fact that the field equations are equivalent with a variational principle has been put into evidence by Martzner and Misner [21], in connection with axially symmetric field. These authors did not arrive at the complex potential, but their method can be applied in our case as follows. Let us suppose that the field is described by the variables (Y^j), for which we have discovered the metric:

$$h_{ij}dY^i dY^j, \tag{3.46}$$

in an ambient space of metric:

$$\gamma_{\alpha\beta}dx^\alpha dx^\beta. \tag{3.47}$$

In this situation, the field equations are derived from a variational principle, connected to the Lagrangian:

$$L = \gamma^{\alpha\beta} h_{ij} \frac{\partial Y^i}{\partial x^\alpha} \frac{\partial Y^j}{\partial x^\beta}. \tag{3.48}$$

In our case, the metric (3.46) is given by (3.42), the field variables being h and $\overline{h}$, or, equivalently, the real and imaginary parts of h.

Therefore, if this variational principle is accepted as a starting point, the main purpose of the gravitational field research would be to produce metrics of the Lobachevsky plane (or related to them), being closely connected to Einstein's field equations, by means of the formalism exposed in the previous paragraph. But, as one can see, such a variational principle exceeds the theoretical frame of employing the Ricci tensor. In fact, one can give examples of its applicability for cases which have nothing to do with Einstein's field equations.

Indeed, the metric of the Lobachevsky plane can be produced as a Calylean metric of an Euclidean plane, for which the absoluteness is a circle with unit radius [9]. This way, the Lobachevsky plane can be put into biunivoc correspondence with the interior side of this circle. The general procedure of metrization of a Calylean space starts with definition of the metric as an anharmonic ratio, being extensively described in many classical volumes (see, *e.g.*, [7]), so that we shall skip the details and give only the final results, which are necessary in our following investigation. Let us suppose that the absoluteness of the space is represented by the quadratic form $\Omega(X, Y)$, where X denotes any vector. The Calylean metric is then given by the differential quadratic form:

$$\frac{-ds^2}{k^2} = \frac{\Omega(dX, dX)}{\Omega(X, X)} - \frac{\Omega^2(X, dX)}{\Omega^2(X, X)}, \tag{3.49}$$

where $\Omega(X, Y)$ is the duplication of $\Omega(X, X)$, and k a constant connected to the space curvature.

In case of the Lobachevsky plane, we have:

$$\begin{aligned} \Omega(X, X) &= 1 - x^2 - y^2, \\ \Omega(X, dX) &= -x dx - y dy, \\ \Omega(dX, dX) &= -dx^2 - dy^2, \end{aligned} \tag{3.50}$$

which yields:

$$\frac{ds^2}{k^2} = \frac{(1 - y^2)dx^2 + 2xy dx dy + (1 - x^2)dy^2}{(1 - x^2 - y^2)^2}. \tag{3.51}$$

Performing now the coordinate transformation (for details, see [2]-[4], [22]-[24]):

$$x = \frac{h\overline{h} - 1}{h\overline{h} + 1}, \quad y = \frac{h + \overline{h}}{h\overline{h} + 1}, \tag{3.52}$$

the metric (3.51) becomes identical to (3.42):

$$\frac{ds^2}{k^2} = -4\frac{dhd\overline{h}}{(h - \overline{h})^2}. \tag{3.53}$$

As easily seen, the absoluteness $1 - x^2 - y^2 = 0$ goes to the straight line $\Im h = 0$, while straight lines of the Euclidean plane go to circles with centres on the real axis of the complex plane (h).

The previous theory is applicable to any case of existence of a real conic, being easy to demonstrate that the Calylean metric attached to this conic is the metric generated by the transformation group which leaves it invariant. Before studying the metric (3.53) in more detail, let us give an example of potential applicability of the variational principle Matzner-Misner regarding this metric. This example has nothing to so with the curvature tensor, because is based of the idea of avoiding the use of this tensor.

W. Rindler [27] showed that the motions performed in a general space-time, on torsion-free trajectories with constant curvature, stand for a natural generalization of the so-called hyperbolic motions (with constant acceleration) of the Minkowskian space-time. These motions are particularly important, since are described in coordinates analogue to Kruskal coordinates [28] and, in addition, by means of these coordinates was put into evidence - for the first time – non-equivalence of the second quantization in Minkowskian frames, on the one hand, and uniformly accelerated frames, on the other [15]. An interesting result of this theory is the fact that an observer engaged in a uniformly accelerated motion in the Minkowskian vacuum is situated in a thermal bath, its temperature being directly proportional to his acceleration [29]. For the one-dimensional case, the equation of Minkowskian hyperbolic motion writes [28]:

$$Z^2 - c^2T^2 = \frac{c^4}{\alpha^2}, \tag{3.54}$$

where α is the magnitude of acceleration, the other quantities possessing the usual significance. The last equation explains the term "hyperbolic" associated with this motion. The quantity:

$$X = \frac{c^2}{\alpha} \tag{3.55}$$

is called "Rindler coordinate", attached to hyperbolic motion. Equation (3.54) then yields:

$$Z^2 = c^2T^2 + X^2, \tag{3.56}$$

giving us possibility of explicitating the rest condition $Z = const.$ for a particle engaged in a hyperbolic motion. Equation (3.56) can be interpreted as absoluteness in coordinates:

$$x = \frac{X}{Z}, \quad y = \frac{cT}{Z}, \tag{3.57}$$

and obtain the metric (3.53) by taking:

$$h = \frac{y \pm i\sqrt{1 - x^2 - y^2}}{1 - x}, \tag{3.58}$$

or, in view of (3.57):

$$h = \frac{cT \pm i\sqrt{Z^2 - c^2T^2 - X^2}}{Z - X}. \tag{3.59}$$

The corresponding Ernst equations, written in the form (3.45) with $h = u + iv$ are equivalent to the pair of real equations:

$$\begin{aligned} &v\nabla^2 u - 2\nabla u \nabla v = 0, \\ &v\nabla^2 v - (\nabla v)^2 + (\nabla u)^2 = 0, \end{aligned} \tag{3.60}$$

where the differential operators are taken in an arbitrary three-dimensional metric. As one observes, $T = 0$ implies $h = iv$, where:

$$v = \frac{\sqrt{Z^2 - X^2}}{Z - X}, \tag{3.61}$$

meaning that the Ernst potential is purely real. In this case, the field is characterized by the second equation (3.60), which is easily solved. If the change of variable $v = e^{\xi}$ is performed, it reduces to Laplace equation:

$$\nabla^2 \xi = 0. \tag{3.62}$$

Let us solve this equation for one of the simplest cases, *i.e.* spherical symmetry of ξ. In spherical coordinates (r, θ, φ), equation (3.62) writes:

$$\frac{1}{r^2}\frac{d}{dr}\left(r^2\frac{d\xi}{dr}\right) = 0,$$

with solution:

$$\xi(r) = C_1 - \frac{C_2}{r}, \quad C_1, C_2 = const. \tag{3.63}$$

or, by the use of (3.61) and an adequate choice of the integration constants:

$$\frac{Z+X}{Z-X} = e^{-\frac{K}{r}}. \tag{3.64}$$

Equation (3.64) can be used to extract coordinate X:

$$X = -Z\tanh\left(\frac{K}{r}\right). \tag{3.65}$$

In the limit $K/r \to 0$, one can write:

$$X = -\frac{ZK}{r},$$

or, by means of (3.55):

$$\alpha = -\kappa^2 r, \tag{3.66}$$

with $\kappa^2 = ZK/c^2$. It then follows that, for $T = 0$, α can be considered as acceleration in a radial oscillatory motion or, more correctly - if passing to the limit leading to (3.66) is considered – as a centripetal acceleration in a uniform circular motion. How to interpret this fact? The variational principle used to obtain the above result emerges from the metric (3.53), which is invariant with respect to a certain group of coordinate rational transformations. If we adopt the equivalence principle [8], then we can say that, for an arbitrary point, the Rindler coordinate X represents the gravitational field intensity, while the mentioned group represents transition between various gravitational fields acting on that point. The appearance of a simultaneous action of these gravitational fields is a motion of rotation characterized by the centripetal force given by (3.66) at the initial (Minkowski!) moment far from the common centre of the field generating sources.

Therefore, by replacing the principle of independence of simultaneous actions, as a linear composition of field intensities at a point, by the *a priori* invariance of fields action with respect to a certain group, one could conceive a theory of gravitation free of the usual inherent contradictions of actual theory [25].

As we have mentioned, the starting point of our previous discussion is not the curvature tensor, but a certain (hyperbolic) motion. But, if similar results, obtained by means of Ernst's equation, that is of the Matzner-Misner variational principle, prove their strength, eliminating some contradictions in the actual theory of gravitation, it means that this variational principle is of a maximum generality and leads to results which can involve the space curvature.

Expressing our belief in the last possibility, next we shall give a detailed description of the metric (3.53), followed by its generalization and, implicitly, generalization of the variational principle generated by it, as well as an attempt of operational definition of the fields constructing the Lagrangian.

3.3 Geodesics and parallel transport in Lobachevsky plane

The metric (3.53) or, equivalently, (3.43), is a conformally Euclidean metric through the factor:

$$E(u, v) = \frac{1}{v^2}. \tag{3.67}$$

Calculation of geodesics of this metric can be performed, without taking into account the Christoffel symbols of the second kind, by using the geodesics of the Euclidean plane (the straight lines), by means of transformation (3.52). Indeed, let:

$$ax + by + c = 0 \tag{3.68}$$

be a straight line in the Euclidean plane. If transformation (3.52) is applied, the straight line becomes the circle:

$$(a + c)h\overline{h} - b(h + \overline{h}) - (a - c) = 0, \tag{3.69}$$

or, by means of $h = u + iv$:

$$(a + c)(u^2 + v^2) + 2bu - (a - c) = 0, \tag{3.70}$$

It can be easily seen that the centre of circle is on u-axis ($v = 0$) and, therefore, the circle is orthogonal to this axis. The axis $v = 0$

becomes now absoluteness of the space. Condition for the straight line (3.68) to intersect the absoluteness $x^2 + y^2 - 1 = 0$ in two real points then writes:

$$a^2 + b^2 - c^2 > 0.$$

This relation is equivalent to condition of intersection between circle (3.70) and $v = 0$ in two real points.

Lobacewski's geometry, as an absolute geometry, is usually limited to the interior part of absoluteness, that is to the superior complex half-plane ($v > 0$).

We consider here an extension of this absolute geometry, taking into account the exterior of the absoluteness as well.

As an alternative procedure, equations (3.70) can be also obtained by means of a variational principle involving the Lagrangian:

$$L = \frac{1}{2}\frac{\dot{u}^2 + \dot{v}^2}{v^2}, \tag{3.71}$$

where $\dot{u}$ and $\dot{v}$ are derivatives of u and v with respect to an arbitrary affine parameter. The Euler-Lagrange equations then write:

$$\begin{aligned} \frac{d^2u}{dt^2} - \frac{2}{v}\frac{du}{dt}\frac{dv}{dt} &= 0, \\ \frac{d^2v}{dt^2} + \frac{\dot{u}^2 - \dot{v}^2}{v} &= 0, \end{aligned} \tag{3.72}$$

and can be easily integrated. As a result, we obtain families of circles with centres on u-axis:

$$(u - \alpha)^2 + v^2 = \beta^2, \tag{3.73}$$

as expected.

Another important result connected to Lobachevsky plane is that it can be applied on pseudo-sphere [9], *i.e.* on a surface with constant negative Gaussian curvature. Indeed, the only non-zero Riemann-Christoffel symbols are [9]:

$$R^1_{212} = -R^2_{112} = -\frac{1}{v^2}, \tag{3.74}$$

so that the curvature invariant R is:

$$R = v^2 R^1_{212} = -1, \tag{3.75}$$

which completes the proof.

Once the metric of the Lobachevsky plane is given, we can define a vector parallel transport in the usual sense of Levi-Civita [5]; the vector origin moves on geodesic, the angle between the vector and the tangent to geodesic at the current point being permanently constant.

Taking advantage of the fact that metric of the plane is conform-Euclidean, one can calculate the angle between the initial and the transported vector by integrating the equation [9]:

$$d\varphi = \frac{1}{2}\left(\frac{\partial(\ln E)}{\partial v}du - \frac{\partial(\ln E)}{\partial u}dv\right) \tag{3.76}$$

along the given curve. Here $E(u, v)$ denotes the conformal factor of the metric which, in our case, is given by (3.67). Introducing (3.67) into (3.76), we are left with:

$$d\varphi = -\frac{du}{v}. \tag{3.77}$$

This formula is important, as we shall see in the following investigation.

3.4 The Barbilian group

As we have mentioned, the group (3.40) is a group in two variables with three parameters. Writing explicitly its transformations so that the identical transformation corresponds to the null values of parameters, we can conveniently write them as:

$$h' = \frac{h + a^1}{a^3 h + a^2 + 1}, \tag{3.78}$$

where a^1, a^2, a^3 are the real group parameters. Next, let us express its infinitesimal transformation [31] – *i.e.* those transformations characterized by infinitesimal values of parameters – and expand (3.78) together with its complex conjugate about the parameters origin ($a^1 = a^2 = a^3 = 0$). The result is

$$\begin{aligned} h' &= h + a^1 - ha^2 - h^2 a^3, \\ \overline{h'} &= \overline{h} + a^1 - \overline{h}a^2 - \overline{h^2} a^3, \end{aligned} \tag{3.79}$$

where, this time, a^1, a^2, a^3 are infinitely small quantities of the first order, while their higher powers have been neglected.

According to (3.79), one and the same transformation $h \to h'$ corresponds to an infinite number of values of parameters a^1, a^2, a^3. This result can be easily verified: solving the system (3.79) with respect to a^1, we realize that it is compatible undetermined. It is said that the group is *multiple transitive*; there is an infinity of its homographies which make it possible transition $h \to h'$.

Recalling significance of h variables as being field potentials, from the physical point of view would be desirable for the group to be *simply transitive*: to a given system of parameters a^1, a^2, a^3 corresponds one transition $(h \to h')$, and only one. This desire is purely classical, and corresponds – for example – to the construction of Maxwell's equations, under given initial conditions.

A necessary condition for a group to be simply transitive is, obviously, that the number of variables equals the number of parameters. In this case, (3.79) should contain three equations, instead of two. If, in addition, the matrix of coefficients of (a^1, a^2, a^3) would be of rank three, then certainly the group would be simply transitive.

Keeping in mind the origin of this group, the only possibility is to add a new field variable (whose significance shall be explained later on). This operation can be done in many ways, under condition of keeping the structure of the starting group. As one can observe by examining (3.79), this condition is given by the infinitesimal operators:

$$
\begin{aligned}
X_1^0 &= \frac{\partial}{\partial h} + \frac{\partial}{\partial \overline{h}}, \\
X_2^0 &= -h\frac{\partial}{\partial h} - \overline{h}\frac{\partial}{\partial \overline{h}}, \\
X_3^0 &= -h^2\frac{\partial}{\partial h} - \overline{h}^2\frac{\partial}{\partial \overline{h}},
\end{aligned}
\tag{3.80}
$$

through the equations:

$$
\begin{aligned}
[X_1^0, X_2^0] &= -X_1^0, \\
[X_2^0, X_3^0] &= -X_3^0, \\
[X_3^0, X_1^0] &= -2X_2^0.
\end{aligned}
\tag{3.81}
$$

Rewriting the infinitesimal transformation (3.79) in three variables as

$$h' = h + a^1 - ha^2 - h^2a^3,$$
$$\overline{h'} = \overline{h} + a^1 - \overline{h}a^2 - \overline{h^2}a^3, \tag{3.82}$$
$$k' = k + \gamma(h, \overline{h}, k)a^3,$$

where k is the new variable introduced due to already mentioned reasons, the first two differential operators of (3.80) do not change, while the third becomes:

$$X_3 = X_3^0 + \gamma\frac{\partial}{\partial k}. \tag{3.83}$$

Imposing to the new operators the same structural relations as those given in (3.81), it follows that γ is defined up to a multiplicative function of the new variable:

$$\gamma(h, \overline{h}, k) = \psi(k)(h - \overline{h}). \tag{3.84}$$

Choosing – for reasons that soon shall become obvious – $\psi(k) = -k$, it follows that the infinitesimal operators of the new simply transitive group (it can be easily verified that the system (3.82) has a unique solution in a^1, a^2, a^3 under condition $h - \overline{h} \neq 0$, $k \neq 0$) are given by:

$$X_1 = X_1^0, \quad X_2 = X_2^0,$$
$$X_3 = -X_3^0 - (h - \overline{h})k\frac{\partial}{\partial k}. \tag{3.85}$$

As it can be observed, k has to be a unimodular complex variable. This can be proved by solving the equation:

$$X_3 f = 0.$$

It is satisfied if and only if:

$$k = \text{const.}\,\frac{\overline{h}}{h}.$$

If the constant is adequately adjusted, we can limit ourselves, indeed, to considering k as a unimodular factor, but without being the ratio between $\overline{h}$ and h. This was the reason for choosing $\psi = -k$ in (3.84): if we set $k = e^{i\varphi}$, then we shall have $k\frac{\partial}{\partial k} = \frac{\partial}{\partial \varphi}$, which means derivative with respect to the phase of k.

As can easily be verified, the infinitesimal relations (3.82) correspond to the finite transformations:

$$k = \frac{h + a^1}{a^3 h + a^2 + 1},$$
$$k' = \frac{a^3 \overline{h} + a^2 + 1}{a^3 h + a^2 + 1} k, \tag{3.86}$$

and this is the Barbilian group, named after Romanian mathematician Dan Barbilian [6]. Here are some properties of this group.

Its structure is given by equations (3.81) and, consequently, the structure constants are:

$$C^1_{12} = C^3_{23} = -1, \quad C^2_{31} = -2, \tag{3.87}$$

the rest of them being zero. Therefore, the invariant quadratic form is given by the "quadratic" tensor of the group [31], that is:

$$C_{\alpha\beta} = C^\mu_{\alpha\nu} C^\nu_{\beta\mu}, \tag{3.88}$$

where summation over repeated indices is understood. Using (3.87) and (3.88), the tensor $(C_{\alpha\beta})$ writes:

$$C_{\alpha\beta} = \begin{pmatrix} 0 & 0 & -4 \\ 0 & 2 & 0 \\ -4 & 0 & 0 \end{pmatrix} \tag{3.89}$$

meaning that the invariant metric of the group has the form:

$$\frac{ds^2}{k_0^2} = \omega_0^2 - 4\omega_1\omega_2, \tag{3.90}$$

where k_0 is an arbitrary factor and (ω_α) three differential 1-forms, absolutely invariant through the group. Barbilian takes these 1-forms as being given by

$$\omega_0 = i \left(\frac{dk}{k} - \frac{dh + d\overline{h}}{h - \overline{h}} \right),$$
$$\omega_1 = \overline{\omega}_2 = \frac{dh}{k(h - \overline{h})}, \tag{3.91}$$

so that the metric becomes:

$$\frac{ds^2}{k_0^2} = -\left(\frac{dk}{k} - \frac{dh + d\overline{h}}{h - \overline{h}} \right)^2 + 4\frac{dh\, d\overline{h}}{(h - \overline{h})^2}. \tag{3.92}$$

It is worthwhile to mention a property connected to the integral geometry: the Barbilian group (3.86) is measurable. Indeed, it is simply transitive and, since its structure vector [32]:

$$C_\alpha = C^\nu_{\nu\alpha} \tag{3.93}$$

is identically null, as it can be seen from (3.87), this means that it possess an invariant function given by:

$$F(h, \overline{h}, k) = -\frac{1}{(h-\overline{h})^2 k}, \tag{3.94}$$

which is the inverse of the modulus of determinant of the linear system (3.82) in unknowns (a^i).

As a result, in space of the field variables $(h.\overline{h}, k)$ can *a priori* be constructed a probabilistic theory [19], based on the elementary probability:

$$dP(h, \overline{h}, k) = -\frac{dh \wedge d\overline{h} \wedge dk}{(h-\overline{h})^2 k}, \tag{3.95}$$

where $\wedge$ denotes the exterior product of the 1-forms.

Finally, it is notably to observe the Barbilian group is "latently" contained in the group of homographic transformations (3.78). Indeed, let us focus our attention on metric (3.92): it reduces to metric (3.53) for $\omega_0 = 0$. By means of the usual relations:

$$h = u + iv, \quad k = e^{i\varphi} \tag{3.96}$$

the first relation (3.91) yields:

$$\omega_0 = -\left(d\varphi + \frac{du}{v}\right), \tag{3.97}$$

and therefore, condition $\omega_0 = 0$ becomes:

$$d\varphi = -\frac{du}{v}, \tag{3.98}$$

which is the definition (3.90) of the Levi-Civita parallelism angle [9]. Therefore, $\omega_0 = 0$ represents the Levi-Civita parallelism of the Lobachevsky plane. This is a theorem established by Barbilian [6].

We further observe that the variational principle Matzner-Misner is based on the Lagrangian given by (3.92):

$$L = 4\frac{\nabla h \nabla \overline{h}}{(h-\overline{h})^2} - \left(\frac{\nabla k}{k} - \frac{\nabla h + \nabla \overline{h}}{h - \overline{h}}\right)^2, \tag{3.99}$$

and, if notations (3.96) are used, this yields:

$$\begin{aligned} &\nabla^2\varphi + \frac{1}{v}\nabla^2 u - \frac{1}{v^2}\nabla u \nabla v = 0, \\ &\nabla^2\varphi - \frac{1}{v}\nabla\varphi\nabla u = 0, \\ &\nabla^2 v - \frac{1}{v}(\nabla v)^2 - \nabla u \nabla\varphi = 0. \end{aligned} \tag{3.100}$$

These equations reduce to (3.96), if the differential condition $\omega_0 = 0$ is accepted under operator condition:

$$\nabla\varphi + \frac{\nabla u}{v} = 0, \tag{3.101}$$

in which case the first equation (3.100) writes:

$$\nabla\left(\nabla\varphi + \frac{\nabla u}{v}\right) = 0, \tag{3.102}$$

being identically satisfied. The last two equations (3.100) can also be transformed by means of (3.101) and (3.102).

In solving equations (3.100) (or their complex equivalent) we meet the same difficulties as in the case of Ernst's equations for the usual approach. In addition, appears a difficulty connected with the unimodular variable k. The only thing we know about it is that, in the limit case of the Poicaré group, it stands for the angle of natural parallelism of the Lobachevsky plane. Due to this fact, in the next chapter we shall formulate some possible definitions of this variable, starting with *a priori* notions on synchronization, frames deformability, and quantum observability associated to a certain frame.

3.5 Correlations of the Barbilian's group with various standard dynamics. Perspectives.

This metric is also typical to a genuine characterization of an ensemble of oscillators, starting from their classical description by the solutions of a second order differential equation.

Indeed, the second order differential equation characterizing a classical damped harmonic oscillator, can be written as

$$M\ddot{q} + 2R\dot{q} + Kq = 0 \tag{3.103}$$

with obvious notation for first and second derivatives of the relevant coordinate q. The solutions of equation (3.104) form a two-dimensional manifold depending on *three arbitrary parameters*. They can be written in the form

$$q(t) = e^{-\lambda t}(he^{i(\omega t+\phi)} + \overline{h}e^{-i(\omega t+\phi)}); \;\; M^2\omega^2 \equiv MK - R^2, \;\; M\lambda \equiv R. \tag{3.104}$$

Therefore the solutions represent an ensemble of oscillators of the same frequency, in which the element is identified by three parameters $h, \overline{h}$ and $e^{i\phi}$. Having the experience of the blackbody radiation, one might say that the frequency is somehow statistically related to this ensemble of oscillators, the same way the temperature is related to the kinetic energy, assuming of course that it is possible to find such a statistics. This statement can even be made more precise from algebraical point of view.

Indeed, the ratio of any two linearly independent solutions of the differential equation (3.103), τ say, is a solution of the following differential equation

$$\{\tau, t\} = 2\omega^2, \tag{3.105}$$

where the curly brackets denote the so-called Schwartz derivative of τ with respect to time, defined by [22]-[24]

$$\{\tau, t\} \equiv \frac{d}{dt}\left(\frac{\ddot{\tau}}{\dot{\tau}}\right) - \frac{1}{2}\left(\frac{\ddot{\tau}}{\dot{\tau}}\right)^2. \tag{3.106}$$

This differential expression, and therefore the left hand side of the equation (3.105) along with it, is invariant with respect to the homographic transformation of the function, i.e. the ratio of two fundamental solutions of the equation (3.103)

$$\tau \leftrightarrow \tau' = \frac{a\tau + b}{c\tau + d} \tag{3.107}$$

with a, b, c, d four real parameters. The set of all transformations (3.107) corresponding to all the possible values of these parameters is obviously the group $SL(2, R)$.

Thus the ensemble of all oscillators of the same frequency is in a one-to-one correspondence with the transformations of $SL(2, R)$. This allows us to construct a 'personal' parameter τ, so to speak, for each oscillator of the ensemble of possible solutions of the classical equation of the harmonic oscillator, guided by the form of the general solution of equation (3.105). This solution can indeed be written as

$$\tau' = u + v\tan(\omega t + \phi), \tag{3.108}$$

where u, v and ϕ are constants, characterizing a given oscillator from the ensemble as before. Identifying the phase from (3.108) with that from (3.104), we can write the personal parameter of an oscillator in the form

$$\tau' = \frac{h + \bar{h}\tau}{1+\tau}, \quad h \equiv= u + iv, \quad \bar{h} = u - iv, \quad \tau \equiv e^{2i(\omega t + \phi)}. \tag{3.109}$$

This equation reveals the Barbilian form of the roots of a cubic equation (see the paragraph 2.10), so our results can be summarized as follows:

i) each oscillator represents a family of cubic equations depending on time;

ii) the initial conditions of the oscillator - the amplitudes - are given by the roots of the Hessian of the cubic family (see (2.165)).

In physical terms this means that the ensemble of oscillators represents a family of matrices giving a measured field. The physical quantities accessible to measurement are the normal and shear components of the field. The time of physical evolution of the field is given by the phase angle of orientation of the octahedral vector in the local octahedral plane (see the paragraph 2.10).

Thus, the ensemble of initial conditions of the oscillators corresponding to the same frequency can be organized as a geometry of the hyperbolic plane in the representation of Poincaré (see (2.152)). Therefore these oscillators correspond to a situation where their initial conditions can be chosen from among the points of a hyperbolic plane. One such situation is that of the classical motion representing an atom. Therefore these oscillators are related to the structure of the nucleus, and with the stresses inside nucleus. In other words, the general constitutive law enforces naturally by the representation of stresses and strains represented as 3×3 matrices has a dynamical interpretation in terms of oscillators.

The metric (3.92) offers now a natural possibility of extension of the harmonic principle that we associated with a Kepler problem, by a functional which in fact represents the strain of the nuclear matter in the general case. The extension amounts to expressing the energy of mapping by the functional

$$E(\mathbf{\Phi}) = \iiint \left\{ -4\frac{\nabla h \cdot \nabla \overline{h}}{(h-\overline{h})^2} + \left(\frac{dk}{k} - \frac{\nabla h + \nabla \overline{h}}{h - \overline{h}} \right) \right\} (d^3\boldsymbol{x}). \quad (3.110)$$

The variational principle would then express this simple fact: the most general strain of the nuclear matter is replicated in the atomic structure by the variation of the eccentricity of the electronic orbits. Likewise, inside nucleus, the variation is replicated by an ensemble of harmonic oscillators of the same frequency. The second term from this functional ([22]) is analogous to the baryonic term from the theory of Skyrme, with the only difference that now the whole functional is homogeneous. In the language of skyrmion technology however ([22]), it still represents hyperbolic skyrmions, because the geometrical character of the problem is dictated by the metric from equation (3.92).

In this form, the theory illustrates the strong ties of the model of nuclear matter with the ideas of confinement of matter in general. In order to show this, we will present now a solution to the variational principle related to the energy functional from equation (3.110). For better intuitive understanding of the solution, let us start with a 'strange' classical dynamics, the one related to the forces involved in the problem of confinement of the classical ideal gas (see [22]), of magnitude inversely proportional with the distance between molecules. This force is also involved in a Newtonian description of the inertia ([22]) which, again should appear as quite natural in view of the common background of the general relativity and Skyrme theory ([22]).

Assuming a Newtonian dynamics for this force, the equations of motion can be written in the form

$$\ddot{\boldsymbol{r}} + \frac{\mu^2}{r^2}\boldsymbol{r} = 0.$$

This is a plane motion, because the force is central. In polar coordinates of the plane of motion, the equations of motion splits into the system of two differential equations

$$\ddot{r} + r\dot{\theta}^2 + \frac{\mu}{r} = 0, \quad r\ddot{\theta} + 2\dot{r}\dot{\theta} = 0. \quad (3.111)$$

The second of these gives the area constant:

$$r^2\dot{\theta} = \dot{a}. \quad (3.112)$$

The first equation (3.111) can then be integrated as follows ([22]): first change the time variable into

$$\tau(t) = \int \frac{dt}{r^2} \therefore \theta'(\tau) = \dot{a}, \tag{3.113}$$

where the prime represents differentiation with respect to τ. Now, if we change the dependent variable into $\xi(\tau) = 1/r$, the first equation (3.111) becomes

$$\xi'' + (\dot{a}^2 - \mu^2)\xi = 0, \quad \xi(\tau) \equiv \frac{1}{r}. \tag{3.114}$$

This is the equation of a harmonic oscillator, with the frequency dictated by the rate of area, leading to either trigonometric or hyperbolic functions. We consider only the first case, when the solutions are of the form

$$\xi(\tau) = A\sin(\Omega\tau) + B\cos(\Omega\tau), \quad \Omega^2 \equiv \dot{a}^2 - \mu^2 > 0. \tag{3.115}$$

Now, we can find the relation between the Newtonian time and the new time τ. Indeed taking equation (3.115) into equation (3.113) gives

$$t \equiv \int \frac{d\tau}{\xi^2(\tau)} = \int \frac{d\tau}{[A\sin(\Omega\tau) + B\cos(\Omega\tau)]^2} \tag{3.116}$$

and this integral leads to ([27])

$$\omega\tau = \phi_0 + \tan^{-1}[(A^2 + B^2)\Omega t], \tag{3.117}$$

where ϕ_0 is a constant of integration. Interestingly enough, equation (3.115) represents a particular Kepler motion, corresponding to null gravitational constant, or null mass, or even null charge as it were. It cannot be therefore interpreted in terms of the motion of a material point around another attracting material point. However it can be interpreted in terms of an abstract kinematics, suggested by the analogy between equation (3.117) and the solution (3.108) of the Schwartzian equation, which can be interpreted as the eigenvalue of a stress matrix.

Consider indeed the kinematics generated by differential forms (2.188). In terms of these the metric (3.92) assumes the Lorentzian form (2.189). Along the geodesics of this metric the rates represented by the differential forms (2.188) are constant, so that we can find those geodesics from some differential equations involving a parameter linear in the arlength from equation (3.92). These are

$$d\phi + \frac{du}{v} = a \cdot dt,$$
$$\cos\phi \frac{du}{v} + \sin\phi \frac{dv}{v} = b \cdot dt, \quad -\sin\phi \frac{du}{v} + \cos\phi \frac{dv}{v} = c \cdot dt, \quad (3.118)$$

with a, b and c some constants. The last two of these equations give

$$\frac{\dot{u}}{v} = b\cos\phi - c\sin\phi, \quad \frac{\dot{v}}{v} = b\sin\phi + c\cos\phi \qquad (3.119)$$

and then from the first of (3.118) we have

$$\dot{\phi} = a - b\cos\phi + c\sin\phi \qquad (3.120)$$

an equation which can be integrated right away. We prefer to perform this integration by putting the right hand side of (3.120) in the form of a perfect square, in order to show that this corresponds to the equation (3.116) above. Indeed, if we take $2\Omega\tau \equiv \phi$, we can write the integral in the form (3.117) for a, b, c given by

$$a \equiv \frac{A^2 + B^2}{2}, \quad b \equiv \frac{A^2 - B^2}{2}, \quad c \equiv AB. \qquad (3.121)$$

This gives an interpretation to the classical 'time' variable τ, provided we know something about this abstract kinematics. Insofar as the parameters u and v are concerned, using equation (3.119) we have

$$\frac{v}{v_0} = a - b\cos\phi + c\sin\phi, \qquad (3.122)$$

where v_0 is another integration constant. Therefore v represents the inverse square of the position vector of the motion previously described. On the other hand, for the parameter u we find the following solution

$$v_0(u - u_0) = b\sin\phi + c\cos\phi, \qquad (3.123)$$

where u_0 is another constant of integration.

Now one can find a particular solution of the variational principle applied to the energy functional (3.110) along the geodesics given by equations (3.122) and (3.123), if we assume that their parameter (and therefore the phase ϕ) is a solution of the Laplace equation. This can be proved, by continuing to work in real parameters u, v

and ϕ as before. The Euler-Lagrange equations associated with the variational principle applied to (3.110) are:

$$\nabla\left(\nabla\phi+\frac{\nabla u}{v}\right)=0,\ \nabla^2\phi-\nabla\phi\cdot\frac{\nabla v}{v}=0,\ \nabla^2(\ln v)-\nabla\phi\frac{\nabla u}{v}=0. \tag{3.124}$$

On the other hand, the geodesics of the metric (3.92) are solutions of the system of differential equations:

$$\left(\frac{u'}{v}\right)'+\phi'\frac{v'}{v}=0,\ \phi''-\phi'\frac{v'}{v}=0,\ (\ln v)''-\phi'\frac{u'}{v}=0, \tag{3.125}$$

where the prime means differentiation with respect to the parameter of geodesics. Now, if the functions u, v and ϕ depend on position only through the parameter of geodesics, the equations (3.124) can be written in the form

$$\left(\phi'+\frac{u'}{v}\right)(\nabla t)^2+\left(\phi'+\frac{u'}{v}\right)\nabla^2 t=0,\ \left(\phi''-\phi'\frac{v'}{v}\right)(\nabla t)^2+\phi'\nabla^2 t=0,$$
$$\left((\ln v)''-\phi'\frac{u'}{v}\right)(\nabla t)^2+(\ln v)'\nabla^2 t=0.$$

The first terms of these equations are zero as a consequence of the equations of geodesics. So we have still another form of the harmonic principle: in fairly general conditions, the harmonic mapping corresponding to energy (3.110) is given by the geodesics of the metric, provided their parameter is a regular harmonic function. This of course makes the phase ϕ the arctangent of such a function, in view of the equation (3.120) above.

Therefore, at least in this particular instance, we have to focus on the geodesics of the metric (3.92). The Killing vectors of the metric represent conservation laws, and thus the parameters a, b, c from equations (3.122) and (3.123) represent expression of these conservation laws. The equations (3.122) and (3.123) themselves represent two Kepler motions, and by this we are certainly in position to know exactly the field of application of the above kinematics: it is the theory of space stresses, involved in the Kepler problem. These stresses are induced in the core of the solar system for instance, or in the nucleus, and they represent a material point in the sense of Hertz whose particles, acted upon by the inverses of the space elastic forces, behave like an ideal gas. The density of this material point varies inversely proportional with the square of distance from the origin of the reference frame, which is actually its origin. We have to insist upon these two aspects of the problem of action at distance, for they are instrumental in understanding how the nucleus works.

It is well-known that one of the first theories of nuclear matter was that of a gas model ([22]). It was not so successful, but certainly touched a fundamental aspect of the problem of structure of nuclear matter, which turns out to be actually the fundamental problem regarding the structure of the matter in general. In view of the discussion above, we may assume the following scenario: the particles of nucleus are decided 'in pairs' by the inversion transformation between external Newtonian forces and the forces, internal to the nucleus, of confinement. These last ones are represented as harmonic oscillators – gluons as it were – and described by a general dynamics. The kinematics of a pair is represented as two Kepler motions given by equations (3.122) and (3.123) above. Here, of course, we assume that the pair is 'accidentally decided' by inversion, but once decided it is described by an actual state of stress whose kinematics can be classically described. This state of stress can be physically characterized by a statistics of the kind specifically connected with the light (for details see [22]).

With the equations (3.122) and (3.123) above, we certainly can extend to matter, particularly to the nucleus, the Mac Cullagh's view about the structure of light (for details, see [22]). Indeed the two equations represent two harmonic oscillators, as well as two Kepler motions. In classical terms, one can say that two particles (partons) inside nucleus have independent motions of a Kepler kind. Even though classically described, such an image should be statistically accessible to measurement through some kind of stresses or strains, and reflected in the eccentricity of electronic orbits.

This image of the structure of the nucleus can be made even more precise, from the very classical point of view we advocate in this work. Indeed, inasmuch as we are thinking in the framework of the Newtonian natural philosophy, the above view on the Hertz's particles the partons of nuclear matter recovers, and solves we should say, in a 'quantum-mechanical' way, one of the most important issues of that philosophy, left behind by Newton in his theory of forces.

We will turn now back to chapter 2, where we have shown that the classical description of the Kepler motion leaves room for an elegant description of the nucleus through the geometry of the hyperbolic plane, in terms of eccentricity and orientation of the Kepler orbit. More to the point, if we express the complex number h, representing the asymptotic direction of the orbit, as a function of the eccentricity e of the orbit and its orientation, ω say, (for details see [22]) then the absolute metric obtained as a result of the constraint which expresses that the Kepler orbit is a closed conic section, is given by (2.155).

The angle ω gives the orientation of the orbit in its plane with respect to an arbitrary direction. If ψ is a regular harmonic function in space, then the complex number h gives a mapping from space to the hyperbolic plane. Therefore, if the nucleus is made of Hertz particles and we identify the harmonic coordinates of these particles with the forces between them ([22]), then the eccentricity of this orbit is a physical expression of the statistics of intranuclear forces between constituent particles of nucleus.

The metric (2.155) is also the metric of a section of hyperbolic space, used in the construction of hyperbolic skyrmions [22]. The form of this metric is, in general

$$(ds)^2 = (d\psi)^2 + \sinh^2\psi(d\theta^2 + \sin^2\theta d\phi^2), \tag{3.126}$$

where θ and ϕ are usual spherical polar angles. Obviously, (2.155) can be obtained from (3.126) if we agree that ω represents the geodesic arc on the unit sphere. But, as we have shown in [27], equation (3.126) is also the absolute metric of the space of relativistic velocities (the Fock metric) and probably there are still many meanings of it, of which one is of a particular cosmological interest.

It is indeed, worth mentioning the fact that, if it is to be applied to the interior of atomic nucleus, the theory of gravitation in its Einsteinian form can only refer to a cosmology. More than this, this cosmology has to be spatially homogeneous, in the sense of existence of a transitive group describing it. Now, inasmuch as this homogeneity involves a three-parameter group, with the structure of Barbilian group, the homogeneous space should be of the Bianchi type VIII in the language developed by cosmologists (for details, see the references from [22]). Consequently the metric from equation (3.92) above, must have some obvious connections with the metric of some spatially homogeneous cosmologies. One such cosmology aiming explicitly to the introduction of the material points in the form of galaxies, is the Gödel's cosmology, taking into consideration the nonzero density of matter by a cosmological term in Einstein's equations (for details see [22]). The metric of such a space time is given by

$$(dx_0 + e^{x_1} dx_2)^2 - (dx_1)^2 - \frac{e^{2x_1}}{2}(dx_2)^2 - (dx_3)^2 \qquad (3.127)$$

with the usual relativistic notation taking x_0 as the time coordinate and $x_{1,2,3}$ as space coordinates. One can see that the section with constant x_3 of this space-time has a metric almost identical with the metric from equation (3.92). In fact it can be written in the form given in equation (3.92) by a transformation which offers 'physical meaning', so to speak, to time and plane coordinates:

$$x_0 = \sqrt{2}\phi, \quad x_1 = \ln\frac{\sqrt{2}}{v}, \quad x_2 = u. \qquad (3.128)$$

Here we used our usual notation, $h \equiv u + iv$, $k \equiv e^{i\phi}$. Rotating cosmologies carry the particular significance of an universe in motion, but with important proviso that the time is given by this motion, it is not a priori parameter. In this specific case, one can say that Gödel cosmology, when applied to 'nuclear universe', gives a dynamic explanation of this universe in terms of the classical motions of its particles. We like to point out that this 'dynamic explanation' can be done in terms of Kepler motions as geodesics of the Barbilian metric-equations (3.122) and (3.123) above. In fact this might justify the point of view that the Barbilian space is a phase space for the Kepler motion in general. In the case of Kepler motion the velocity vector follows a circle while the moving body follows the conic describing the motion ([22]). One can assume that, in general, the dynamics of the motion is described in a phase space and the trajectories are represented by two conics. If the equation (3.122) represents a conic:

$$\frac{1}{r} = a - b\cos\phi + c\sin\phi, \qquad (3.129)$$

then this radial motion has a speed given by

$$\dot{r} = b\sin\phi + c\cos\phi, \qquad (3.130)$$

in view of the time transformation from equation (3.113). Here the time derivative of the radial coordinate is taken with respect to the very time of the motion, as it should be. Eliminating the angle between equations (3.129) and (3.130), one can even get a conservation law. This simple fact shows that, the radial component of the Kepler motion is actually of the same nature – at least from an algebraic point of view – with the free radial motion ([22]).

And while we are on the subject, another aspect of the problem of analogy transcending the scale deserves special mention. The metric (3.126) caught attention of the specialists in relation to the cosmic background radiation ([22]). There are indications of space topologies in the temperature correlations of the cosmic background radiation, that point out toward finite topology of space. These cannot be realized but only within hyperbolic cosmologies. But there is more to this interpretation of the hyperbolic metric, even outside the Hamiltonian theory: it touches the very essence of the classical Newtonian theory of forces. In hindsight it may be taken as a mark of the conservation of force ([22]).

The metric from equation (3.126) can be brought to the form (3.92), for both are metric of constant negative curvature. As shown above, one just needs for this a transformation offering physical meaning to coordinates. Now, the fact that metric (3.126) was used in describing the skyrmions with zero mass pions, incites us in constructing for it a fundamental skyrmion with distinguished significance in the very roots of Newtonian mechanics. For that skyrmion, the equation (3.121) - and therefore equation (3.92) - still maintain the meaning of absolute metric ([22]), but this time it is also related to a conservation law. Let us follow closely the logic of such a construction, starting even from the Newtonian theory of forces.

Fact is that Newtons definition for the central force, the kind of force needed in astronomical researches as well as in the theory of nuclear particles, is actually a definition based on the concept of measurement of forces. We stressed this idea quite a few times in the present work. It amounts basically to the fact that the ratio of magnitudes of forces acting on a planet in different directions from the plane of motion, but determining the very same Keplerian orbit, covered in the same time interval, can be recognized in the elements of that orbit ([22]). This is the essential point of the Corollary 3 of the Proposition VII from Principia, which contains the most general definition of the gravitational force, leading to the idea of its 'universality'.

There is, however, an essential point left uncovered by Newton, and this is the problem of mass. He insisted on the proportionality of the central force of heavens with the product of the masses involved in interaction, which is quite an arbitrary assumption. It is not wrong, by any means – the history proves it – but, we should say, incomplete. It was obvious even to Newton himself that in order to be logically correct, the hypothesis was quite insufficient as it stood, and needed to be amended with 'perturbation' terms in order to account for the fact that the Sun is not fixed, and we do not deal here with material points without space expanse (see for details Popescu, 1988). Nevertheless, used as Newton gave it first, the theory of forces led in a natural way to the equation of Poisson ([22]), which further led to the general relativity (see also [22] on the subject of 'natural way').

Newton's principle of measurement compares the forces along two different lines from the plane of motion of the planet, in a ratio depending only on the geometrical elements of the motion ([22]). In the long run, one might say, let us stress it once more, that his philosophy is simply the one saying that the ratios of the forces, acting upon a planet in any two directions in plane, are to be read in the planet's motion. The force must be always central, otherwise in the vector model of forces the orbit is no longer plane. In view of todays astronomical knowledge this seems to be quite an ideal speculative conclusion, for no motion in the universe seems to be plane. However, in the times of Newton the only real thing to be taken into consideration was the Kepler's synthesis of planetary observations, which plainly sustained the conclusion of plane motion. This is why we take special precaution of talking of Keplerian orbits, whenever we have to discuss the subject of Newtonian forces: they are related exclusively to the Keplerian setup of the geometry of motion.

However, Newton's hypothesis about masses in the expression of the magnitude of forces, aims to describe, and introduce within the concept of central force, not what happens along two different lines of action in plane, but what happens along the same line of action. In view of the reciprocity of gravitational interaction, the action at distance is the same no matter of the point of view along its line of action, and it should therefore be characterized by the magnitude of the vector of force. Moreover, by the very same token, this magnitude should not depend on the direction of action along the same line. The product of masses in the expresion of force is an algebraical monomial satisfying this requirement, but so is the sum, or any kind of average for that matter, like for instance the reduced mass. Later on, even the equations of motion of the classical dynamics led to the conclusion that the masses are only convenient coefficients, and their choice is, in the long run, only justified by the simplicity of mathematical description [22].

As we see it, the point at issue is very probably the fact that the third principle of dynamics is no more actual over the space. If the forces do not act in the same point and, moreover if, physically speaking, this point does not remain unchanged by their action, the forces cannot be compared. In other words, the point where forces act in keeping with the third principle should be a Hertz particle. This is hardly the case with the matter in general, but even if it would be such a case the central action still needs to be amended. Indeed, the quantity of matter must be essential. However, one cannot say that a planet perceives in the Sun the same amount of matter that the Sun itself perceives in the planet. In other words, the fact that the masses involved in interaction are different should be inherently contained in the definition of the very magnitude of forces [22].

Then this part of the definition of forces should be taken out of the classical vector formalism, which apparently does not allow for a proper accomodation of the concept of mass. And it can be taken indeed, by means of a quantum definition of the measurement, in the axiomatic manner in which the isospin is introduced in the theory of nuclear forces, or the spin is introduced in quantum mechanics. Namely, the situation just described is represented by the following hermitian 2×2 matrix depending explicitly on the direction in space – vector characteristic – but having two eigenvalues equal in magnitude, independent of that direction:

$$\boldsymbol{Q} \equiv \begin{pmatrix} \cos\theta & \sin\theta e^{i\phi} \\ \sin\theta e^{-i\phi} & -\cos\theta \end{pmatrix}. \tag{3.131}$$

This matrix has, indeed, the eigenvalues ± 1, or any two real numbers equal in magnitude and opposite in sign for that matter. Consequently it can represent the masses of two ideal particles along any direction in space, acting on each other with a central Newtonian force. Based on matrix (3.131) we can build an ansatz. First notice that any 2×2 matrix of the form

$$\boldsymbol{M} = \lambda \boldsymbol{E} + \mu \boldsymbol{Q}, \tag{3.132}$$

where λ and μ are real, and $\boldsymbol{E}$ is the 2×2 identity matrix, has two different eigenvalues not depending on the direction. These are $(\lambda \pm \mu)$. The ansatz is then exactly the Skyrme ansatz, inasmuch as equation (3.132) can be written in the form

$$\boldsymbol{M} = \exp(\psi \boldsymbol{Q}). \tag{3.133}$$

The matrix $\boldsymbol{M}$ then represents indeed a skyrmion, because it is referring to quantities that characterize the nucleus. First, the absolute metric of the matrices (3.132) or (3.133) is just the metric from equation (3.126), where we only need to take

$$\frac{\mu}{\lambda} = \tanh\phi. \tag{3.134}$$

What this representation tells us is that the measurement provides the masses $(\lambda \pm \mu)$ of two particles in interaction, which they perceive into each other through this interaction, and which are independent of the direction $(\theta, \pm\phi)$ of the straight line joining them. The formula (3.134) is then justified by the inequality of the masses of the two partners involved in interaction. But this does not mean that the masses themselves are independent of the environment of the particles inside nucleus. Indeed, in order to represent the nuclear matter, ψ needs to be a solution of the Laplace equation over the space occupied by the nucleus. This certainly brings analytical dependence of the ratio of masses on the whole environment of the line joining the two particles. Moreover, by our discussion of the forces from [22], ψ must be a force.

It is hard to see a situation illustrating this idea in the macroworld. However, the Newtonian suggestion of the analogy transcending the scale, and the importance of the homogeneous universes with the metric (3.126), revealed by the analysis of the cosmic background radiation, seem to point toward the old thesis of the conservation of force. For, in the microworld the Skyrme ansatz is a common situation, so to speak. The history of physics plainly illustrates it, which is why we sustain that the Skyrme ansatz is actually only a natural finish of the classical natural philosophy of forces. Indeed, take the case of proton and electron. In the Rutherford atom, described as a classical Kepler motion, they have significantly different masses. In this case, we can certainly say that the value of ψ in equation (3.134) is very small, the forces are elastic, and so the metric in equation (3.126) is very nearly Euclidean. This is why the atom could be described as a Kepler motion, or even as a harmonic oscillator in the first place. This description is legitimate even from the point of view of the isospin proper. Indeed, it naturally contains the limit case of quantities equal in magnitude and opposite in sign, that the electron and the proton perceive into one another in special occasions. This is, of course, the case of their charges which, as well-known, are equal in magnitude and of the opposed signs. The pairs of equal mass, for instance the proton and antiproton, or the electron and positron, are here represented by the limit of very large ψ, i.e. of the very small difference of their masses.

The theory of nuclear forces asks necessarily for a kind of algebraic theory of masses, whereby the idea of direction intimately enters the theoretical implement (for instance in the form of mixing angles). And the above approach shows plainly how the notion of direction should work, and where the ratio of masses intervenes. However this idea of richness of the notion of direction, and of its close relation with the ratio of masses is not new, by any means. With specific reference to the quantum theory of fundamental Newtonian structures and their relation with cosmology and cosmogony, the idea was elaborated to a large extent by Eddington (for details see [22]). Our only addition here is perhaps that the quantum theory has a direct descent from the classical Newtonian theory of forces, which it just completes and finishes naturally. The Skyrme theory seems to be indeed, the only right addition to that classical theory, in the sense of completing it in its very own terms.

References

1. Adler R., Bazin, M., Schiffer M.: Introduction to General Relativity, Mc Graw Hill, (1972).
2. Agop, M., Gavriluţ, A., Păun, V.P., Filipeanu, D., Luca, F.A., Grecea, C., Topliceanu, L.: Entropy, 18, 160: DOI: 10.3390/e1805016((2016).
3. Agop M., Gavriluţ A.: Reports on Mathematical Physics, 2, 76, 231, (2015).
4. Agop M., Gavril S., Gavriluţ A., Rezuş E.: International Journal of Modern Physics B, 29, 1, (2015).
5. Arnold, V.I., Mathematical Methods of Classical Mechanics, Springer, (1989).
6. Barbilian, D.: Opera Mathematica, Romanian Academy Publishing House, vol.I, Bucharest, (1967).

7. Barbilian, D.: Opera Didactica, Technical Publishing House, Bucharest, Vol.I, II, III, (1968), (1971), (1974).

8. Einstein, A.: Relativity: The Special and the General Theory, Barnes ans Noble Library of Essential Reading, (2004).

9. Efimov, N.: Geometrie Superieure, Mir, Moscou, (1981).

10. Ernst, F.J.: Physical Review, 167, 1175, (1968).

11. Ernst, F.J.: Physical Review, 168, 1415, (1968).

12. Ernst, F.J.: J. Math. Phys., 12, 239, (1971).

13. Fock, V.A.: The Theory of Space, Time and Gravitation, Macmillan, (1964).

14. Fré, R.: Acta Phys. Polonica, B, 10, 383, (1979).

15. Fulling, S.A.: Physical Review, D, 7, 2850, (1973).

16. Gottlieb I., Agop M., Jarcău M.: Chaos, Solitons and Fractals, 19, 705, (2004).

17. Ionescu-Pallas, N.: General Relativity and Cosmology, (in Romanian), Scientific and Encyclopedic Publishing House, Bucharest, (1980).

18. Israel, W., Wilson, G.A.: J. Math. Phys., 13, 865, (1972).

19. Jaynes, E.T.: Found. Phys., 3, 477, (1973).

20. Kramer, D. et al. (E Schmutzer, Editor), Exact solutions of Einsteins field equations, Cambridge University Press, (1980).

21. Matzner, R.A., Misner, C.W.: Physical Review, 154, 1229, (1967).

22. Mazilu N., Agop, M.: Skyrmions. A Great Finishing Touch to Classical Newtonian Philosophy, New York: Nova Science Publishers, (2012).

23. Mazilu, N., Agop, M.: La răscrucea teoriilor. Intre Newton şi Einstein - Universul Barbilian, Editura Ars Longa, Iaşi, (2010).

24. Mercheş, I., Agop, M.: Differentiability and Fractality in Dynamics of Physical Systems, World Scientific, Singapore, (2016).

25. Popescu, I.N.: Gravitation, Scientific and Encyclopedic Publishing House, Bucharest, (1982).

26. Ray, J.R.: Wei, M.S., Nuovo Cimento, B, 42, 151, (1977).

27. Rindler, W.: Physical Review, 119, 2082, (1960).

28. Rindler, W.: Amer. J. Phys., 34, 1174, (1966).

29. Unruh, W.G.: Physical Review, D, 14, 870, (1976).

30. Vysotsky, M.I.: Iader. Fiz., 43, 1338, (1986).

31. Vranceanu, G.: Lectures on differential geometry, Vol. III, Didactic and Pedagogical Publishing House, Bucharest, (1979).

32. Vranceanu, G., Filipescu, D.: Elements of integral geometry, Romanian Academy Publishing House, Bucharest, (1982).

CHAPTER 4

Frames, measurements and field variables

4.1 The notion of a frame

Both experimental and theoretical physics pay tribute to the notion of frame. Position and/or motion of a body can only be described with respect to a certain frame. From the experimental point of view, the frame is given by a system of bodies – or, in a more abstract manner – of points relatively fixed with respect to each other that intermediate measurements of distances and directions. Interference of the speculative thinking made then possible to detach the notion of coordinates of a point in a given frame: spatially, the point is localized by three real numbers – its coordinates – and, if it moves, also by the time moment determined by an horology attached to the point when its localization is determined (for details see [1]-[16]).

Theoretically speaking, the events or phenomena are described in coordinates chosen as conveniently as possible. Usually, the coordinate systems are chosen so as to satisfy a certain type of symmetry, or to offer possibility of a direct, experimental testing of theoretical investigation (for details see [17]-[30]).

In principle, calculations can be performed in any coordinate system, since there are many transformation formulas from one system to another. This reciprocal transformability led to detachment of the notion of frame from its initial, experimental meaning. A frame usually used in geometry is the frame of the vectors tangent to the coordinate lines (the so-called natural frame [13]). Due to the fact that transitions between such frames are given as derivatives of some well-known functions, these transformations are called *holonomic*. As compared to these purely mathematical frames, the physical frame is marked by a high degree of arbitrarity: not always its abstract defining elements – the basic vectors – can define tangent to them coordinate lines. As a result, transformations between two such frames cannot be expressed by usual functions, and the basis of such frames is called *unholonomic* [10]. Nevertheless, these frames are commonly used in experimental physics, the holonomic frames being only abstractions necessary for calculations, as we shall here in after show.

From the physical point of view, a coordinate system has to correspond to reality, being measurable or interpreted in terms of measurable quantities. But, in almost all cases the coordinate are not measurable. In astronomy, for example, position of a star can be determined either by Cartesian coordinates, or spherical coordinates: azimuth, height, and distance to the star. In both cases we have to do with quantities whose significance or – alternatively – the effectively determined value are unknown. For example, the Cartesian coordinates cannot be measured, even if the space is supposed to be absolutely Euclidean. In astronomy, one usually employs the second set of coordinates, but the radial distance – obtained by physical means – is quantitatively affected up to a large scale of arbitrariness. Even qualitatively, if the space is non-euclidean, the distance changes its initial significance, while the angles do not indicate the real direction, but only an apparent, local one. Anyhow, these frames – or parts of them, which eventually could play the role of a "theodolite" – keep their significance, which ignores *displacements* and take into account only *rotations*.

These facts are important, since among all notions used in physics, the notion of direction is somewhat transcendental. For example, the experimental device has to obey with good precision the classical laws. In other words, the difference between classical laws (applicable to measurement instruments) and quantum laws (concerning micro-objects), implying the scale of measurement of mass, length, and time extends on quantities implying *displacements*, but not on *directions*, implying rotations only. There are two aspects here we net to point out. The first concerns the intuitive fact that a direction – no matter how it is settled – can serve as a frame for both microscopic and macroscopic phenomena. The second refers to the fact that the uncertainty relations, implying angular variables and associated operators, demand precautions in their definition [8]; this could display a certain "classicism" of the considered variables.

Indiscernibility of the coordinate systems, together with the fact that in measurements only "unholonomic" physical frames are used, clearly show the necessity of reporting calculations to the last ones. Besides, only these frames are used, for example, in *operational* definition of the space-time metric elements. However, as observed by N. Ionescu Pallas [11], even models of universe defined by ingenious operational procedures, such as Milne's kinematic model, can be reformulated by means of the notion of metric. There exists a firm basis for metric theories: Einstein's general relativity. What exactly is missing in physics (or, more limited, in cosmology) to overcome the actual *status quo* is the physical interpretation – *i.e.* in a physical picture – of the coordinates defining the metric. In other words, if a system of coordinates can usually define a frame (of the tangent vectors), the actual problem is inverse: given a frame, find the compatible coordinates. The previous theory of complex potential (see Chapter 2) allows us to come close to solution to this problem, since becomes clear that the only "holonomic" coordinates belong to the "ambient" three-dimensional metric, which can be a working hypothesis.

Therefore, these coordinates are connected to the measured quantities (*e.g.* a potential) only by a certain function, as solution of a partial differential equation. But this fact implies extension of the operational procedures beyond the usual ones: measurement of distances, even by light signals, and horologies synchronization, in the same way. These procedures are intrinsically connected, and, since synchronization is marked by a high degree of arbitrariness, the definition of distance has the same property (for more details, see [29]).

4.2 Measurement process and the field variables

One way of defining a measurement procedure connected to the physical frame, as previously accepted, consists in a speculation based on quantum measurability, that is on Hermitian operators whose proper values (eigenvalues) stand for measured values of the quantities with attached operators. To explain such a point of view, we first have to accept that on a frame can be settled directions, indifferently from our working phenomenological limits (micro- or macrocosmos). Concretely, we admit that – artificially (with rational intervention) or intrinsically – a physical frame is endowed with "teodolitic devices" whose orientation is frame-independent and, most of all, independent of the way the frame is considered (*e.g.* independent of orientation of the three attached Cartesian axes). Regarding the teodolitic device, we have to realize that inside this structure some interactions take place, due to the fields caught from the vision direction; by means of these interactions, some quantities connected to this field are experimentally determined. Due to its nature (construction), the device measurements do not depend on the direction of observation, or parameters associated to this direction. In one word, the *internal* process of measurement is independent of direction.

Under these circumstances, a teodolit must be characterized by a Hermitic operator which can depend on orientation, but having eigenvalues independent of direction. Such an operator can be constructed in case of *two* real and distinct eigenvalues. If, for example, these eigenvalues are ± 1, the operator looks like [16]:

$$Q = \begin{vmatrix} \cos\theta & \sin\theta e^{i\varphi} \\ \sin\theta e^{-i\varphi} & -\cos\theta \end{vmatrix}, \tag{4.1}$$

where θ and φ are the azimuth and height of the direction, or, better to say, its parameters on unit sphere. The two eigenvectors of the matrix Q are defined up to an arbitrary phase factor, but the ratios of their components, which are the fixed points of the matrix Q considered as homography:

$$z_1 = \cot\frac{\theta}{2}e^{i\varphi}, \quad z_2 = -\tan\frac{\theta}{2}e^{i\varphi} \tag{4.2}$$

are in biunivocal correspondence with the considered direction. These parameters (z_1, more frequently) serve to express direction as a point function on the sphere of unit radius, considering this sphere as an one sheet hyperboloid of imaginary generatrices [2]. The matrix:

$$M = xE + yQ, \tag{4.3}$$

where x and y are real and E is the 2×2 unit matrix, has the same fixed points as Q, but the eigenvalues:

$$\lambda_1 = x + y, \quad \lambda_2 = x - y, \tag{4.4}$$

are direction-independent. Consequently, the matrix (4.3) which is obviously Hermitic, satisfy our demands to characterize a teodolit and we shall use it in the investigation to follow. It then remains to put into evidence the physical significance of the eigenvalues λ_1 and λ_2 or, better, of x and y.

Parameters x and y have a certain "internal" freedom, which must be considered. Indeed, the result of measurements on some direction reflects a "sum" of influences of the same nature *from all spatial directions* that do not explicitly interfere. The mentioned "sum" (for example, superposition of gravitational and/or electromagnetic fields) depends on the physical structure of the teodolit and could be expressed as a certain *internal* freedom concerning the measurement process if we consider it – for example – as being given by a linear group of transformations in x and y.

It is very difficult, obviously, to *a priori* put into evidence such a group, but we shall try to do it by analogy with some known cases. To this end, let us consider the similitude transformations of matrix M:

$$M' = L^{-1}\, M\, R, \tag{4.5}$$

where L and R are two 2×2 matrices. Obviously, such transformations correlate matrices M and M', each of them with its own characteristics for different directions and different eigenvalues. Since these matrices affect directions, they are not of much help, but – as we shall see – they can be used for some analogy.

As we have pointed out in the previous chapter, obtaining of a Cayleyen metric could be useful in many cases, due to its connection with the Matzner-Misner variational principle. Since x and y can be considered as measured field variables, we shall repeat our attempt of obtaining such an invariant, but this time with respect to transformations (4.5). In this case, the absoluteness is given by the matrices M of null-determinant [2], wich signifies that one or both eigenvalues are zero. By virtue of previous observations, such matrices characterize the device while no "measurement" – in no direction – is performed, that is the cosmic background, Procedure of the metric construction was given in section 3.2, with the result:

$$-\frac{ds^2}{k^2} = \frac{dx^2 - dy^2}{x^2 - y^2} - \frac{(xdx - ydy)^2}{(x^2 - y^2)^2} - \frac{y^2}{x^2 - y^2} d\Omega^2, \tag{4.6}$$

where:

$$d\Omega^2 = d\theta^2 + \sin^2\theta\, d\varphi^2$$

is the metric on unit sphere, and k is an arbitrary constant.

It is worthwhile to mention that, if transformation (4.5) takes place with preservation of the determinant of matrix M $(L = R)$:

$$det\, M = x^2 - y^2 = const.,$$

the metric (4.6) becomes:

$$-(x^2 - y^2)ds^2 = dx^2 - dy^2 - y^2 d\Omega^2, \tag{4.7}$$

and looks very similar to the metric of Minkowski space, if we take $x = ct,\ y = r$, where t is the time, and r the radial coordinate.

But even ignoring such a case, it can be shown that x and y obey some transformations which are equivalent to Lorentz transformations. One can observe that metric (4.6) is dependent only on the ratio x/y. Indeed, by means of transformation:

$$x = X \cosh T, \qquad y = X \sinh T, \tag{4.8}$$

(4.6) can be written as:

$$\frac{ds^2}{k^2} = dT^2 + \sinh^2 T\, d\Omega^2, \tag{4.9}$$

which proves the statement. Therefore, no matter how the coordinate X transforms, the metric is totally indifferent.

This metric has a special importance: it is the "public space" metric or, formally speaking, the spacial part of N.Ionescu Pallas's [11] stationary hyperbolic metric. This fact is expected, since the metric (4.7) is *conformally* Minkowskian, whose conformity factor is obtained as benefit only of the fact that it is an absolute metric.

Therefore, using (4.9) as a spatial metric relatively to Ernst equation, we shall solve this equation for the case of a real potential, independently of the angles. In this case, Ernst's equation [9] reduces to:

$$\frac{1}{\sinh^2 T}(\sinh^2 T\, f')' = (f)^2.$$

where the accent denotes the derivative with respect to T, with the solution:

$$f(T) = f_0 e^{-\alpha \cosh T},$$

leading to the hyperbolic stationary metrics [23]:

$$ds^2 = f_0 e^{-a \cosh T} dt^2 - \frac{1}{f_0} e^{a \cosh T}(dT^2 + \sinh^2 T\, d\Omega^2), \tag{4.10}$$

t being the temporal coordinate. Matrices (4.10) stand for a generalization of that given by N.Ionescu Pallas [11], obtained from (4.10) with $f_0 = 1$, $a = 0$.

We are now interested in coordinate transformations (X, T); since such transformations are connected to transformations in x and y, it is important to set the problem of their finding. To this end, we shall employ a very much used viewpoint in modern theoretical physics, namely considering vectors as abstract quantities satisfying the quaternion algebra [6]. Indeed, it is obvious that (4.3) can be used to generally express M in terms of Pauli matrices, while the linear transformations in x and y (only) induce transformations of these matrices, which form an absolute orthogonal frame. Basis of such a frame is usually given by three 2×2 mutually anticommutative matrices, say I, J, K, which satisfy the multiplication table:

$$I^2 = J^2 = K^2 = E, \qquad I \cdot J \cdot K = iE$$
$$IJ = -JI = iK, \quad JK = -KJ = iI, \quad KI = -IK = iJ, \tag{4.11}$$

where $i = \sqrt{-1}$. The "vectors" of this space are 2×2 matrices with null trace (involutions), whose norm given by the scalar product:

$$(A \cdot B) = \frac{1}{2}(A.B + B.A)$$

is a multiple of the unit matrix:

$$(M\ M) = (M_1^2 + M_2^2 + M_3^2)E. \tag{4.12}$$

Among frames (4.11), the one given by Pauli's matrices:

$$I_0 = \begin{bmatrix} 0 & i \\ -i & 0 \end{bmatrix}, \quad J_0 = \begin{bmatrix} 0 & 1 \\ 1 & 0 \end{bmatrix}, \quad K_0 = \begin{bmatrix} 1 & 0 \\ 0 & -1 \end{bmatrix} \tag{4.13}$$

plays an essential role in the algebraic theory of spin 1/2.

In such frames, the 2×2 matrices are represented by linear combinations of the unit matrix and the elements (4.11) of the frame. The vector part of the matrix is then given as a linear combination of the last ones only. A null vector does not also mean the null matrix. But, by all means, one or both of its eigenvalues have to be zero; according to the canons of quantum mechanics, the physical significance of measurements is based on these elements. These eigenvalues can also be given as – for example – some functions of x

and y, which are obviously zero. Therefore, in such an extension of the theory, the eigenvalues are not necessarily given by simple linear expressions as (4.4).

If we accept the null vectors as being the "cosmic background", that is the "zero" of the device, it follows their major importance in vacuum problems, such as of the matrices previously discussed. Geometry of the null vectors was detaily studied by Sobczyk [17], and we shall use two of his important theorems.

Let us define in the frame (4.13) a special null vector, associated with direction of I_0, whose canonical form is:

$$N_0 = (E + J_0).K_0 = K_0 + iI_0. \tag{4.14}$$

We are interested in those frame transformations that keep this canonical form, that is those transformations $(I_0, J_0, K_0) \to (I, J, K)$ which maintain the form of the special null vector associated with direction of I:

$$N = (E + J).K = K + iI. \tag{4.15}$$

Sobczyk showed that, in this case, there must exist the transformation:

$$N = e^{\frac{u}{2}N}\, N_0\, e^{-\frac{u}{2}N}, \tag{4.16}$$

where u is a parameter. Writing out explicitly this transformation by means of (4.14) and (4.15), one easily finds the frame transformations that keep the special form of the null vector:

$$\begin{aligned} I &= \left(1 + \frac{u^2}{2}\right) I_0 + iuJ_0 - i\frac{u^2}{2}K_0, \\ J &= -iuI_0 + J_0 - uK_0, \\ K &= -i\frac{u^2}{2}I_0 + uJ_0 + \left(1 - \frac{u^2}{2}\right) K_0. \end{aligned} \tag{4.17}$$

Another Sobczyk's theorem shows that *any* null vector of a frame can be written in terms of the special null vector (4.15) in the form:

$$N' = B\,N\,B, \tag{4.18}$$

where B is a vector orthogonal to direction associated with N:

$$B = UJ + VK. \tag{4.19}$$

The form (4.18) is the so-called Cartan-Whittaker representation of the null vector. In view of (4.19), it can be written as:

$$N' = x_1 I + x_2 J + x_3 K, \tag{4.20}$$

where the components:

$$\begin{aligned} x_1 &= -i(U^2 + V^2), \\ x_2 &= 2\,U\,V, \\ x_3 &= V^2 - U^2 \end{aligned} \tag{4.21}$$

are defined only up to an arbitrary factor and to a permutation which takes into account the direction of the frame associated with the special null vector.

In relation to transformations (4.17), the quantities (4.21) compose the one-parameter linear group:

$$\begin{aligned} x_1' &= \left(1 + \frac{u^2}{2}\right) x_1 - iux_2 - i\frac{u^2}{2} x_3, \\ x_2' &= iux_1 + x_2 + ux_3, \\ x_3' &= -i\frac{u^2}{2} x_1 - ux_2 + \left(1 - \frac{u^2}{2}\right) x_3 \end{aligned} \tag{4.22}$$

which is equivalent to transformations:

$$\begin{aligned} V' &= V, \\ U' &= U + u \end{aligned} \tag{4.23}$$

of the parameters U and V.

Now it is obvious that, if X and T occurring in formula (4.8) are the components of a spinor (4.19): $X = V, T = U$, then transformation (4.23), in parameters x and y, is expressed by:

$$\begin{aligned} x' &= x \cosh\, u + y \sinh\, u, \\ y' &= x \sinh\, u + y \cosh\, u. \end{aligned} \tag{4.24}$$

Transformation (4.23) was found by Rindler [15] while studying analogy between Kruskal coordinates and the uniformly accelerated motion in Minkowski's universe. Hence, (4.24) is equally connected with Minkowski's coordinates, on the one side, and with Rindler's and Kruskal's coordinates, on the other, and represents one-parameter transformations of the abstract frames that make the form (4.15) of special null vectors an invariant. Here we observe that the Kruskal coordinates can be conceived as components of a null vector.

This reality generates some promising analogies but, unfortunately, we cannot go further with coordinates X and T, because T is non-dimensional and, furthermore, the null-vectors (4.21) constructed with them would be summations of expressions which are dimensionally inhomogeneous. The truth is that one of coordinates can be adjusted by an adequate factor so as to be dimensionally correct (as it usually does [11]), but this fact implies an additional arbitrariness. In this case, transformation (4.24) would be defined up to an arbitrary factor, being in fact a 2-parameter transformation. Our problem is, therefore, to directly obtain a generalization for more parameters of the group (4.24), considering x and y as the U and V components of the spinors (4.19), in which case the components of the null vectors (4.21) become dimensionally homogeneous.

At the same time, we also look for a generalization of the group (4.22). To obtain it, one observes that transformations (4.22) leave unchanged the quadratic form:

$$x_1^2 + x_2^2 + x_3^2 = 0, \tag{4.25}$$

which gives the norm of the null vectors. Considering the formal identity between (4.24) and the Euclidean norm, it is naturally to suppose that the most general invariance group is the group with three rotation parameters, whose infinitesimal generators are:

$$\begin{aligned} M_1 &= -i\left(x_2\frac{\partial}{\partial x_3} - x_3\frac{\partial}{\partial x_2}\right), \\ M_2 &= -i\left(x_3\frac{\partial}{\partial x_1} - x_1\frac{\partial}{\partial x_3}\right), \\ M_3 &= -i\left(x_1\frac{\partial}{\partial x_2} - x_2\frac{\partial}{\partial x_1}\right), \end{aligned} \tag{4.26}$$

that really happens. This fact was put into evidence by Yamamoto [20], who observed that the form (4.25) of infinitesimal generators is equivalent to the form resulting from it for:

$$x_1 = \rho\,\sin\omega, \quad x_2 = -\rho\,\cos\omega, \quad x_3 = -i\rho, \tag{4.27}$$

that is:

$$\begin{aligned} M_1 &= \cos\omega\,\rho\frac{\partial}{\partial\rho} - \sin\omega\frac{\partial}{\partial\omega}, \\ M_2 &= \sin\omega\,\rho\frac{\partial}{\partial\rho} + \cos\omega\frac{\partial}{\partial\omega}, \\ M_3 &= -i\frac{\partial}{\partial\omega}. \end{aligned} \tag{4.28}$$

The action of the operators (4.27) on the spin "eigenfunctions":

$$v_+ = \sqrt{\rho}e^{\frac{i\omega}{2}}, \quad v_- = \sqrt{\rho}e^{\frac{-i\omega}{2}} \tag{4.29}$$

reproduces the action of the Pauli matrices, according to relations:

$$\begin{bmatrix} M_1 v_+ \\ M_1 v_- \end{bmatrix} = \frac{1}{2}\begin{bmatrix} 0 & 1 \\ 1 & 0 \end{bmatrix}\begin{bmatrix} v_+ \\ v_- \end{bmatrix},$$

$$\begin{aligned} \begin{bmatrix} M_2 v_+ \\ M_2 v_- \end{bmatrix} &= \frac{1}{2}\begin{bmatrix} 0 & i \\ -i & 0 \end{bmatrix}\begin{bmatrix} v_+ \\ v_- \end{bmatrix}, \\ \begin{bmatrix} M_3 v_+ \\ M_3 v_- \end{bmatrix} &= \frac{1}{2}\begin{bmatrix} -1 & 0 \\ 0 & 1 \end{bmatrix}\begin{bmatrix} v_+ \\ v_- \end{bmatrix}. \end{aligned} \tag{4.30}$$

It can be shown in a straight manner that the operators (4.28) satisfy the same algebras as Pauli's matrices.

The infinitesimal transformations (4.28) and (4.26) don't tell us very much about this group. To this end, we shall look for some finite transformations, generated by the infinitesimal ones. In this respect, we shall set operators (4.28) in a form capable to put into evidence its isomorphism to Barbilian group, considered in the previous chapter. The new operators are given by the linear combinations:

$$\begin{aligned} X_1 &= M_1 - iM_2, \\ X_2 &= M_3, \\ X_3 &= -(M_1 + iM_2), \end{aligned} \tag{4.31}$$

and satisfy the same structure relations as the infinitesimal generators of the Barbilian group.

Taking α and $\overline{\alpha}$ as group variables, with $\alpha = v_+$ from (4.29), the operators (4.31) also write:

$$Y_1 = \overline{\alpha}\frac{\partial}{\partial\overline{\alpha}}, \quad Y_2 = \frac{1}{2}\left(\alpha\frac{\partial}{\partial\alpha} - \overline{\alpha}\frac{\partial}{\partial\overline{\alpha}}\right), \quad Y_3 = -\alpha\frac{\partial}{\partial\alpha}. \tag{4.32}$$

In order to find out the finite transformations generated by these infinitesimal transformations, we shall proceed to determine invariant functions of the operator [19]:

$$U = \mu Y_1 + \nu Y_2 + \lambda Y_3, \quad \mu, \nu, \lambda - \textit{constants},$$

which are solutions of the equation $U\psi = 0$:

$$\left(\mu\bar{\alpha} + \frac{\nu}{2}\alpha\right)\frac{\partial\psi}{\partial\alpha} - \left(\frac{\nu}{2}\bar{\alpha} + \lambda\alpha\right)\frac{\partial\psi}{\partial\bar{\alpha}} = 0, \tag{4.33}$$

with the characteristic system:

$$\frac{d\alpha}{\mu\bar{\alpha} + \dfrac{\nu}{2}\alpha} = -\frac{d\bar{\alpha}}{\lambda\alpha + \dfrac{\nu}{2}\alpha} . \tag{4.34}$$

We now look for another two constants m and n, so that the linear combination resulting from (4.34):

$$\frac{m d\alpha - n d\bar{\alpha}}{m\left(\dfrac{\nu}{2}\alpha + \mu\bar{\alpha}\right) - n\left(\lambda\alpha + \dfrac{\nu}{2}\bar{\alpha}\right)} = d\tau \tag{4.35}$$

is a total differential. This can happen only under conditions:

$$\begin{aligned} \left(\frac{\nu}{2} - \rho\right) m - \lambda n = 0, \\ \mu m - \left(\frac{\nu}{2} - \rho\right) n = 0, \end{aligned} \tag{4.36}$$

where ρ is an arbitrary factor. This system is compatible only for the following values of ρ:

$$\rho_1 = \frac{\nu}{2} + \sqrt{\lambda\mu}, \quad \rho_2 = \frac{\nu}{2} - \sqrt{\lambda\mu}.$$

For the first value, (4.35) yields the integral:

$$\sqrt{\lambda}\alpha + \sqrt{\mu}\,\bar{\alpha} = (\sqrt{\lambda}\alpha_0 + \sqrt{\mu}\bar{\alpha}_0)e^{\left(\frac{\nu}{2} + \sqrt{\lambda\mu}\right)\tau}, \tag{4.37}$$

and for the second, the integral:

$$\sqrt{\lambda}\alpha - \sqrt{\mu}\,\bar{\alpha} = (\sqrt{\lambda}\alpha_0 - \sqrt{\mu}\bar{\alpha}_0)e^{\left(\frac{\nu}{2} - \sqrt{\lambda\mu}\right)\tau}, \tag{4.38}$$

where α_0 and $\overline{\alpha}_0$ are the values of α and $\overline{\alpha}$ for $\tau = 0$. Relations (4.36) and (4.37) give:

$$\alpha = e^{\frac{\nu}{2}}\left(\alpha_0 \cosh\sqrt{\lambda\mu}\tau + \sqrt{\frac{\mu}{\lambda}}\overline{\alpha}_0 \sinh\sqrt{\lambda\mu}\tau\right),$$
$$\overline{\alpha} = e^{\frac{\nu}{2}}\left(\alpha_0\sqrt{\frac{\lambda}{\mu}} \sinh\sqrt{\lambda\mu}\tau + \overline{\alpha}_0 \cosh\sqrt{\lambda\mu}\tau\right). \qquad (4.39)$$

In order that (4.39) make sense, it is necessary that λ be the complex conjugate of μ: $\overline{\lambda} = \mu = \gamma e^{i\phi}$, while ν has to be real – and we take it as zero. Therefore, the finite transformations of the group (4.31), generated by the infinitesimal transformations, are given by the *unimodular* group:

$$\alpha = \alpha_0 \cosh\gamma\tau + \overline{\alpha}_0 e^{i\phi} \sinh\gamma\tau,$$
$$\overline{\alpha} = \alpha_0 e^{-i\phi} \sinh\gamma\tau + \overline{\alpha}_0 \cosh\gamma\tau. \qquad (4.40)$$

It is no accident that we have denoted by α the group variable. This notation is commonly used to denote the complex amplitude of the harmonic oscillators, that is the eigenvalue of the annihilation operator in the second quantization [5]. Transformation (4.40) was first given by Stoler [18], in connection with generalization of the eigenstates of the annihilation operator (coherent states). If α_0 describes, for example, an oscillator in a state satisfying the minimum conditions of momentum-coordinate uncertainty, the state α is characterized by the uncertainty relation:

$$(\Delta p)^2(\Delta q)^2 = \frac{1}{4}(1 + \sinh^2\gamma\tau \sin^2\phi), \qquad (4.41)$$

where (Δp) and (Δq) are variations of the momentum, – respectively, coordinate – of the oscillator, while the Planck constant has been taken as equal to unity.

Consequently, the variable α can be considered as complex amplitude of a bosonic field. The fact that the group (4.31) is isomorphic with the Barbilian group shows that the last one is equally entitled to refer to the variables connected with such a field, and this fact – as we shall prove - really happens.

The group induced by (4.40) on variables x and y can be obtained from (4.27), by means of identifications:

$$x_1 = 2xy,$$
$$x_2 = y^2 - x^2, \tag{4.42}$$
$$x_3 = -i(x^2 + y^2)$$

leading to:

$$x = \sqrt{\rho}\cos\frac{\omega}{2},$$
$$y = \sqrt{\rho}\sin\frac{\omega}{2}. \tag{4.43}$$

As one can see, $\alpha_0 = x + iy$, and introducing $\alpha' = x' + iy'$ into (4.40), one obtains:

$$x' = (\cosh\gamma\tau + \sinh\gamma\tau\cos\phi)\,x + \sinh\gamma\tau\sin\phi\, y$$
$$y' = \sinh\gamma\tau\sin\phi\, x + (\cosh\gamma\tau - \sinh\gamma\tau\cos\phi)\,y. \tag{4.44}$$

These transformations reduce to transformations (4.24) for $\phi = \frac{\pi}{2}$ and $\gamma\tau = u$, but the transformations induced on components Cartan-Whittaker (4.42) do not reduce to those given by relations (4.22).

We are particularly interested in physical significance of spinorial variables $(\alpha, \overline{\alpha})$: each of them signifies a bosonic field. For details concerning the second quantization, Stoler's transformation, and generalized coherent states, the reader is invited to consult [22]. For the moment, we realize that treatment of the angular variables on a level with measured coordinates *straightly* leads to spatial significance of the last ones, through the stationary hyperbolic metric of N.Ionescu Pallas, while making abstraction of angular variables and considering only an "internal" group, the measured quantities signify some field amplitudes.

4.3 Significances of the Barbilian group. Perspectives.

The original de Broglie's theory on wave-corpuscule duality is based on a theorem which is implicitly contained in Lorentz transformation [7]. This theorem connects the cyclic frequency of the local horologies, in each point of a spatial domain, with the progressive wave frequency in phase with the horologe. This wave shows the distribution of the oscillators phases on the considered spatial domain [7].

Let us now show that such a distribution – in the true sense of the word – could be given without appealing to Lorentz transformation.

The de Broglie's idea that all horologies have the same cyclic frequency could be materialized through a periodical field which – as well-known – is describable by means of local oscillators. The result discussed in the previous paragraph regarding the isomorphism between the Barbilian group and Stoler's centro-affine unimodular group shows, as we have mentioned, that the Barbilian group could be successfully applied to the field complex amplitudes.

To prove this idea, we use the harmonic oscillator equation:

$$x'' + \omega^2 x = 0, \tag{4.45}$$

where x is the field relevant coordinate and ω the cyclic frequency. This equation stands for a model for any harmonic of the field. The most general solution of (4.45) depends on three constants and can be written as:

$$x(t) = he^{i(\omega t+\varphi)} + \overline{h}e^{-i(\omega t+\varphi)}, \tag{4.46}$$

where the complex amplitude h is most important in the coherent theory of bosonic fields [5]. Quantities h and $\overline{h}$ display the initial conditions, that differ from one spatial point to another. In other words, at a given moment of time, the oscillators situated in various points of the space are in different quantum states and have different phases. There is a problem which logically appears: is it possible to give an aprioric connection between the parameters h and $\overline{h}$, on the one hand, and $e^{i(\omega t+\varphi)}$, on the other, characterizing various oscillators, at a certain moment of time? Since (4.46) is a solution of (4.35), the answer is affirmative: formula (4.45) possesses a "hidden" symmetry, expressed by the homographic group: the ratio $\tau(t)$ of two solutions of equation (4.45) is a solution of Schwartz's equation [13]:

$$\left(\frac{\tau''}{\tau'}\right) - \frac{1}{2}\left(\frac{\tau''}{\tau'}\right)^2 = 2\omega^2. \tag{4.47}$$

This equation is invariant with respect to a homographic transformation of $\tau(t)$: any homographic function of τ is, in its turn, a solution of equation (4.47). Since homography is characteristic for projectivity on a straight line, we can affirm that the ratio of two solutions of equation (4.45) is a projective parameter for the multitude of oscillators of the same frequency in a given spatial domain.

Now, one can easily construct a convenient projective parameter, so as to be in biunivoc correspondence with the oscillator. In this respect, we first observe a "universal projective parameter", as being the ratio of the fundamental solutions of (4.45) (for details, see [24]-[29]):

$$k = e^{2i(\omega t+\varphi)}. \tag{4.48}$$

Any homographic function of this ratio stands for a new projective parameter. One of these functions, namely:

$$\tau(t) = \frac{h+\bar{h}k}{1+k}, \tag{4.49}$$

has the advantage of being specific to each oscillator. In addition, let us consider another function:

$$\tau' = \frac{h'+\bar{h}'k'}{1+k'}, \tag{4.50}$$

which is specific to another oscillator. Since (4.49) and (4.50) are solutions of equation (4.47), between them exists the homographic relation:

$$\tau' = \frac{a\tau+b}{c\tau+d}.$$

Written explicitly, this relations lead to Barbilian group equations:

$$\begin{aligned} h' &= \frac{ah+b}{ch+d}, \\ \bar{h}' &= \frac{a\bar{h}+b}{c\bar{h}+d}, \\ k' &= \frac{c\bar{h}+d}{ch+d}k. \end{aligned} \tag{4.51}$$

It is now clearly understood, through the methodology of Matzner-Misner variational principle, the mechanism of generation of metric of the stationary field. If we describe this field by amplitude and phase of its harmonics at a certain moment of time, then these oscillators form assemblies of a given frequency, which are transitivity varieties of the Barbilian group (de Broglie coherence in Barbilian's sense). The invariant metric of this group generates the Lagrangian leading to Ernst's equations [7].

It is important to mention, in this context, that there also exists a transitivity group between assemblies of *various frequencies*: we meet, this way, the Stoler group, discussed in the previous paragraph. This fact is easily observed (see [22]), since the parameter $(\gamma\tau)$ of the Stoler transformation is precisely the frequencies ratio, when the creation and annihilation operators refer to a harmonic oscillator.

We are now capable to give the *equivalent* of the de Broglie wave function, that is the distribution of phases and amplitudes on an assembly of given frequency. As we observed in the previous chapter, the Barbilian group is measurable, with the elementary measure given by (2.94):

$$dP = -\frac{dh \wedge d\overline{h} \wedge dk}{(h - \overline{h})^2 k}. \tag{4.52}$$

Therefore, the repartition density of such an elementary probability is the invariant function of the group:

$$\frac{1}{(h - \overline{h})^2 k}. \tag{4.53}$$

As usual, the square root of this function is defined up to an arbitrary unimodular factor, and can be considered as analogous to the wave function. But, this time, it does not satisfy the Schrödinger equation, but some equations deriving from Ernst's equations (2.31). Nevertheless, taking into account these equations, one can deduce an important fact: the "wave" function:

$$\psi = \frac{1}{v} e^{i\phi/2} \tag{4.54}$$

obtained from (4.53), satisfies the stationary "Schrödinger equation":

$$\nabla^2 \psi = \left\{ \left(\frac{\nabla v}{v} \right)^2 + \frac{\nabla u \nabla v}{v} - \frac{(\nabla \phi)^2}{4} + \frac{1}{2} i \frac{\nabla \phi \nabla v}{v} \right\} \psi, \tag{4.55}$$

with a complex "eigenvalue". This becomes real if either the oscillator assembly has the same (spatial) phase, or the same amplitude. In the first case, which is very interesting, equation (4.55) reduces to:

$$\nabla^2 \psi = \left(\frac{\nabla v}{v} \right)^2 \psi, \tag{4.56}$$

its "eigenvalue" being the squared "momentum" responsible for coordinate-momentum uncertainty in the hydrodynamic model of quantum mechanics. In fact, both Ernst's equations (2.31) and the way we define the wave function show that a hydrodynamic model of such a "quantum mechanics" is more easily approached [21].

This investigation offers an example for the way how the *a priori* generated probabilities, as measures of some parametric groups, can interfere in fundamental physical theories [12]. Due to importance of Ernst's equation, the Barbilian group seems to be essential in explaining connection between Relativity and Ondulatory Mechanics.

We cannot ignore the fact that the Barbilian group was first defined as a covariant group of the binary cubic forms [3]. In our context, this fact can have a special significance.

Keeping in mind that a horologies has to be attached to any physical frame, each horology is at the same time an oscillator (periodical motions), so that it is clear that the Barbilian group can be used for synchronization. Let us consider a spatial region endowed with "standard" horologies, which are circular motions with unique frequency ω. Since the way of considering the frame is arbitrary, we choose all frames by Cartesian axes, under condition that the trajectory of circular motion representing the horology is in the plane:

$$x + y + z = 0. \tag{4.57}$$

The oscillator motion is then equivalent to three oscillatory motions on the axes, having equal amplitudes and being dephased from each other by $\pi/3$. The projective parameters (4.49) of this motion are solutions of a certain cubic equation, given by Barbilian [3].

However, it is known that – in general – three solutions of a cubic variable can be considered as spatial coordinates [13]: they are the so-called generalized elliptic coordinates. In such coordinates, the variables $h, \overline{h}, k$ are expressed rationally, at the most algebraically. Maybe, this is the hidden reason that, in these coordinates, the Laplace equation always admits separation of variables.

Among significances of the Barbilian group, we cannot neglect the fact that it was put into evidence for the first time as a covariance group of binary cubic forms. This property, in addition to those associated with propagation and synchronization, is overhelming by its implications. Indeed, the *physical* frame is given as a system of points which are *relatively fixed* one from another. They also have to be fixed – for example – with respect to the fields propagation velocities inside the frame (that is why we underline "relatively fixed"). For this reason, definition of the frame must not contain the displacement mobility. In our opinion, this fact embeds the profound significance of de Broglie's theorem [7] regarding the equivalence between the rectilinear motion and propagation of a wave: anything implying displacement can be described by a field variable. This way, we arrive at the qualitative graduation of possible motions of the points of a frame: there are motions that cannot be described by field variables, *i.e.* extremely slow with respect to the temporal scale of processes described in the frame. We can speak, therefore, about *frame strains.*

For example, the galaxy taken as the frame for phenomena of the solar system is straining, and considering this strain one can explain many negative aspects of the gravitation theory [14], [29]. Practically, we discuss here about a gravitational stress tensor and correspondingly about gravitational strain tensor. More that, the notion of frame can also be used in microcosmos: the crystalline net of a metal is a physical frame for the internal processes of the metal (for example, the motion of electrons), but it is straining *proprio motu.* Even if no force acts on the crystal, the crystalline net is straining by fluaj.

Let us make explicit one of the above statements.

The gravitational stress tensor can be associated to a 3×3 matrix (α_{ij}). Then, the characteristic equation of matrix (α_{ij}) is a cubic equation of the form:

$$x^3 - I_1x^2 + I_2x - I_3 = 0, \tag{4.58}$$

where:

$$I_1 = TR(\alpha) = \alpha_{11} + \alpha_{22} + \alpha_{33}, \tag{4.59}$$

$$I_2 = \begin{vmatrix} \alpha_{11} & \alpha_{12} \\ \alpha_{21} & \alpha_{22} \end{vmatrix} + \begin{vmatrix} \alpha_{11} & \alpha_{13} \\ \alpha_{31} & \alpha_{33} \end{vmatrix} + \begin{vmatrix} \alpha_{22} & \alpha_{23} \\ \alpha_{32} & \alpha_{33} \end{vmatrix},$$
$$I_3 = \det(\alpha_{ij}).$$

The solutions x_1, x_2, x_3 of equation (4.58) are the eigenvalues of matrix (α_{ij}), that is the values on the main diagonal after reducing matrix to diagonal form, which is always possible by means of a nonsingular matrix. If the roots x_1, x_2, x_3 are given, this nonsingular matrix can be found.

There are several possibilities to determine the roots of cubic equation (4.58). The most frequently used is the Tartaglia-Cartan procedure, useful for setting solutions on a trigonometric form, as we shall further realize. In order to use this method, it is necessary to bring (4.58) to a form in which the quadratic term is absent. To this end, let us write x as:

$$x = z + h,$$

and determine h in such a way that the term in x^2 disappears. Introducing this form of x into (4.58) and grouping terms with the same power of z, we obtain:

$$z^3 + (3h - I_1)z^2 + (3h^2 - 2I_1h + I_2)z + h^3 - I_1h^2 + I_2h - I_3 = 0.$$

As one can see, the term in z^2 disappears for the choise:

$$h = \frac{1}{3}I_1 = \frac{1}{3}a_1. \tag{4.60}$$

The other coefficients of equation in z, for the same value of h, can be found according to:

$$3h^2 - 2I_1h + I_2 = \frac{1}{3}(3I_2 - I_1^2) \equiv a_2, \tag{4.61}$$
$$h^3 - I_1h^2 + I_2h - I_3 = \frac{1}{3}(-3I_3 + I_1I_2 - \frac{2}{9}I_1^3) \equiv \frac{1}{3}a_3,$$

and equation in z becomes:

$$z^3 - a_2z + \frac{1}{3}a_3 = 0. \tag{4.62}$$

This is the characteristic equation of the deviator of matrix $(\alpha)_{ij}$. Explicitating the invariants a_1, a_2 in terms of the matrix elements α_{ij}, we obtain:

$$\begin{aligned} a_2 = \frac{1}{6}\{(\alpha_{22} - \alpha_{33})^2 + (\alpha_{33} - \alpha_{11})^2 + (\alpha_{11} - \alpha_{22})^2 \\ \frac{3}{2}(\alpha_{12}\alpha_{21} + \alpha_{13}\alpha_{31} + \alpha_{23}\alpha_{32})\}, \\ a_3 = 3\det(\frac{1}{3}a_1\delta_{ij} - \alpha_{ij}). \end{aligned} \tag{4.63}$$

These formulas can be written in terms of the roots x_1, x_2, x_3 of the cubic equation (4.58):

$$a_2 = \frac{1}{6}[(x_2 - x_3)^2 + (x_3 - x_1)^2 + (x_1 - x_2)^2], \tag{4.64}$$
$$a_3 = -\frac{1}{9}[(2x_1 - x_2 - x_3)(2x_2 - x_1 - x_3)(2x_3 - x_1 - x_2)].$$

As one observes, the quantities $(2x_1 - x_2 - x_3)$ etc. are the deviator roots.

To find the roots of the cubic equation (4.62), we look for z under the form:

$$z = \sqrt[3]{p} + \sqrt[3]{q}, \tag{4.65}$$

where the cubic root signifies here a symbol; for example, $\sqrt[3]{p}$ has three values: $1 \cdot |\sqrt[3]{p}|$, $\omega \sqrt[3]{p}$, $\omega^2 |\sqrt[3]{p}|$, where $1, \omega, \omega^2$ are the cube roots of unity:

$$1 + \omega + \omega^2 = 0.$$

The last relation leads to representation:

$$\omega = \exp\left\{i\frac{2\pi}{3}\right\}, \quad \omega^2 = \exp\left\{i\frac{3\pi}{3}\right\} = \exp\left\{-i\frac{2\pi}{3}\right\}. \tag{4.66}$$

Raising equation (4.65) to third power, we are left with:

$$z^3 - 3\sqrt[3]{pq}z - (p + q) = 0. \tag{4.67}$$

Comparing (4.67) with (4.62), we obtain:

$$p + q = -\frac{1}{3}a_3, \quad p \cdot q = \frac{1}{27}a_2^3,$$

allowing to construct the quadratic equation:

$$y^2 + \frac{1}{3}a_3 y + \frac{1}{27}a_2^3 = 0,$$

with the solutions:

$$p = \frac{1}{6}\left\{-a_3 + \frac{1}{\sqrt{3}}\sqrt{3a_3^2 - 4a_2^3}\right\}, \tag{4.68}$$
$$q = \frac{1}{6}\left\{-a_3 - \frac{1}{\sqrt{3}}\sqrt{3a_3^2 - 4a_2^3}\right\}.$$

The quantity under the square root in (4.68):

$$3a_3^2 - 4a_2^3 = \Delta$$

is called *discriminant* of the cubic equation (4.58) or (4.62). It is, up to a factor, the product of the squared differences of the roots of (4.58) – or, obviously, (4.62). Since matrix (α_{ij}) signifies either the stress matrix σ_{ij}, or the strain matrix ε_{ij}, we demand that equation

(4.58) has real roots. To this end, an algebraic theorem says that Δ has to be negative:

$$\Delta = -3k^2.$$

Under these circumstances, relations (4.68) become:

$$p = \frac{1}{6}(-a_3 + ik), \quad q = \frac{1}{6}(-a_3 - ik). \tag{4.69}$$

These relation can be written in a trigonometric form by taking:

$$-a_3 = 6R\cos\varphi, \quad k = 6R\sin\varphi.$$

Thus:

$$R = \left(\frac{a_2}{3}\right)^{3/2} \tag{4.70}$$

and relations (4.69) become:

$$p = \left(\frac{a_2}{3}\right)^{3/2} e^{i\varphi}, \quad q = \left(\frac{a_2}{3}\right)^{3/2} e^{-i\varphi} \tag{4.71}$$

with:

$$\cos\varphi = -\frac{\sqrt{3}}{2}\frac{a_3}{a_2^{3/2}}. \tag{4.72}$$

To fo further, let us choose φ in such way that the forthcoming formulas are written as simple as possible. Such a useful choice belongs to Novozhilov [30] and consists in:

$$\varphi = 3\theta + \frac{\pi}{2}.$$

Relation (4.72) then becomes:

$$\sin 3\theta = \frac{\sqrt{3}}{2} \cdot \frac{a_3}{a_2^{3/2}}, \tag{4.73}$$

so that the roots of the cubic equations (4.58) write:

$$\begin{aligned}
z_1 &= \sqrt[3]{p} + \sqrt[3]{q} = \frac{2}{\sqrt{3}}\sqrt{a_2}\sin 3\theta, \\
z_2 &= \omega\sqrt[3]{p} + \omega^2\sqrt[3]{q} = \frac{2}{\sqrt{3}}\sqrt{a_2}\sin\left(3\theta + \frac{2\pi}{3}\right), \\
z_3 &= \overline{z}_2 = \frac{2}{\sqrt{3}}\sqrt{a_2}\sin\left(3\theta + \frac{4\pi}{3}\right),
\end{aligned} \tag{4.74}$$

and, for the cubic equation (4.58) or (4.62):

$$x_1 = \frac{2}{\sqrt{3}}\sqrt{a_2}\sin 3\theta + \frac{1}{3}a_1,$$
$$x_2 = \frac{2}{\sqrt{3}}\sqrt{a_2}\sin\left(3\theta + \frac{2\pi}{3}\right) + \frac{1}{3}a_1, \tag{4.75}$$
$$x_3 = \frac{2}{\sqrt{3}}\sqrt{a_2}\sin\left(3\theta + \frac{4\pi}{3}\right) + \frac{1}{3}a_1.$$

According to Novozhiloov's designation, θ is the representation angle of matrix (α) [30]. We note that, in the strain theory θ it is connected with the ratio of radii of Mohr's circles, expressing strain of stresses, being characterized by the Lode parameter μ_α [30]:

$$\mu_\alpha = \frac{2x_1 - x_2 - x_3}{x_2 - x_3} = \sqrt{3}\tan 3\theta. \tag{4.76}$$

The previous theory regarding obtaining the eigenvalues of a matrix is very general: it does not take into account the matrix symmetry, its tensor properties, etc. However, if gravitational stresses and gravitational strain are present, we have to consider noth their symmetry and – obviously – tensor properties. The last property is not of much help, but the first somewhat simplifies the problem of gravitational constitutive relations. For each of matrices (ε_{ij}) and (σ_{ij}) one can construct a theory analogous to that developed above. Thus, let us denote by e_1, e_2, e_3 and, respectively, s_1, s_2, s_3 the invariants of these matrices and by ψ, ζ their representation angles.

The basis of gravitational constitutive relations for a gravitational ideal elastic material structure is given by the existence of the gravitational elastic potential, considered as a function of the gravitational strain invariants E_1, E_2, E_3 replacing I_1, I_2, I_3 given above and signifying the gravitational elementary mechanical work. Denoting this function by $\Phi(E_1, E_2, E_3)$, we have:

$$\sigma_{ij} = \frac{\partial \Phi}{\partial \varepsilon_{ij}} \tag{4.77}$$

or:

$$\sigma_{ij} = \sum_{k=1}^{3} \frac{\partial \Phi}{\partial E_k}\frac{\partial E_k}{\partial \varepsilon_{ij}}. \tag{4.78}$$

The derivatives $\frac{\partial \Phi}{\partial E_k}$ play the role of gravitational generalized elastic moduli. To this purpose let us explicitate relations (4.78). It is not difficult to obtain:

$$\begin{aligned}
\sigma_{11} &= \frac{\partial\Phi}{\partial E_1} + \frac{\partial\Phi}{\partial E_2}(\varepsilon_{22}+\varepsilon_{33}) + \frac{\partial\Phi}{\partial E_3}\left(\varepsilon_{22}\varepsilon_{33} - \frac{1}{4}\varepsilon_{23}^2\right),\\
\sigma_{22} &= \frac{\partial\Phi}{\partial E_1} + \frac{\partial\Phi}{\partial E_2}(\varepsilon_{11}+\varepsilon_{22}) + \frac{\partial\Phi}{\partial E_3}\left(\varepsilon_{11}\varepsilon_{33} - \frac{1}{4}\varepsilon_{13}^2\right),\\
\sigma_{33} &= \frac{\partial\Phi}{\partial E_1} + \frac{\partial\Phi}{\partial E_2}(\varepsilon_{11}+\varepsilon_{22}) + \frac{\partial\Phi}{\partial E_3}\left(\varepsilon_{11}\varepsilon_{22} - \frac{1}{4}\varepsilon_{12}^2\right), \qquad (4.79)\\
\sigma_{12} &= -\frac{1}{2}\left\{\frac{\partial\Phi}{\partial E_2}\varepsilon_{12} - \frac{\partial\Phi}{\partial E_3}\left(\frac{1}{2}\varepsilon_{13}\varepsilon_{23} - \varepsilon_{33}\varepsilon_{12}\right)\right\},\\
\sigma_{13} &= -\frac{1}{2}\left\{\frac{\partial\Phi}{\partial E_2}\varepsilon_{13} - \frac{\partial\Phi}{\partial E_3}\left(\frac{1}{2}\varepsilon_{23}\varepsilon_{12} - \varepsilon_{22}\varepsilon_{13}\right)\right\},\\
\sigma_{13} &= -\frac{1}{2}\left\{\frac{\partial\Phi}{\partial E_2}\varepsilon_{23} - \frac{\partial\Phi}{\partial E_3}\left(\frac{1}{2}\varepsilon_{12}\varepsilon_{13} - \varepsilon_{11}\varepsilon_{23}\right)\right\}.
\end{aligned}$$

Now we are able to express s_1, s_2, s_3 in terms of e_1, e_2, e_3. As a first step, let us write s_1, s_2, s_3 by means of E_1, E_2, E_3 :

$$\begin{aligned}
s_1 &= 3\frac{\partial\Phi}{\partial E_1} + 2\frac{\partial\Phi}{\partial E_2}E_1 + \frac{\partial\Phi}{\partial E_3}E_2,\\
s_2 &= \frac{1}{3}\left\{(E_3^2 - 2E_2)\left(\frac{\partial\Phi}{\partial E_2}\right)^2 + (E_1E_2 - 9E_3)\frac{\partial\Phi}{\partial E_2}\frac{\partial\Phi}{\partial E_3} + \right.\\
&\left. +(E_2^2 - 3E_1E_2)\left(\frac{\partial\Phi}{\partial E_3}\right)^2\right\}, \qquad (4.80)\\
s_3 &= \frac{1}{9}\left\{(2E_1^3 - 9E_1E_2 + 27E_3)\left(\frac{\partial\Phi}{\partial E_2}\right)^2 + \right.\\
&+ (2E_1^2E_2 - 18E_2^2 + 27E_1E_3)\left(\frac{\partial\Phi}{\partial E_2}\right)^2\frac{\partial\Phi}{\partial E_3} +\\
&+ (18E_1^2E_3 - 3E_1E_2^2 - 27E_2E_3)\left(\frac{\partial\Phi}{\partial E_3}\right)^2\frac{\partial\Phi}{\partial E_2} +\\
&\left. +(-2E_2^3 + 9E_1E_2E_3 - 27E_3^2)\left(\frac{\partial\Phi}{\partial E_3}\right)^2\right\}.
\end{aligned}$$

Let us now replace derivatives with respect to E_1, E_2, E_3 with derivatives with respect to e_1, e_2, e_3. The result is:

$$\begin{aligned}
\frac{\partial \Phi}{\partial E_1} &= \frac{\partial \Phi}{\partial e_1} + \frac{2}{3}\frac{\partial \Phi}{\partial e_2}E_1 + \frac{\partial \Phi}{\partial e_3}\left(E_2 - \frac{2}{3}E_1^2\right), \\
\frac{\partial \Phi}{\partial E_2} &= -\frac{\partial \Phi}{\partial e_2} + \frac{\partial \Phi}{\partial e_3}E_1, \\
\frac{\partial \Phi}{\partial E_3} &= -3\frac{\partial \Phi}{\partial e_3}.
\end{aligned} \tag{4.81}$$

Introducing (4.81) into (4.80) we are left with:

$$\begin{aligned}
s_1 &= 3\frac{\partial \Phi}{\partial e_1}, \\
s_2 &= e_2\left(\frac{\partial \Phi}{\partial e_2}\right)^2 + 3e_3\left(\frac{\partial \Phi}{\partial e_2}\frac{\partial \Phi}{\partial e_3}\right) + 3e_2^2\left(\frac{\partial \Phi}{\partial e_3}\right)^2, \\
s_3 &= e_3\left(\frac{\partial \Phi}{\partial e_2}\right)^3 + 6e_2^2\left(\frac{\partial \Phi}{\partial e_2}\right)^2\left(\frac{\partial \Phi}{\partial e_3}\right) + \\
&\quad + 9e_2e_3\left(\frac{\partial \Phi}{\partial e_2}\right)\left(\frac{\partial \Phi}{\partial e_3}\right)^2 + 3(3e_3^2 - 2e_2^3)\left(\frac{\partial \Phi}{\partial e_3}\right)^2.
\end{aligned} \tag{4.82}$$

These formulas can be expressed in terms of e_1, e_2, ψ, if one takes into account that:

$$\sin 3\psi = \frac{\sqrt{3}}{2}\frac{e_3}{e_2^{3/2}} \equiv \alpha.$$

We then have:

$$\begin{aligned}
s_1 &= 3\frac{\partial \Phi}{\partial e_1}, \\
s_2 &= e_2\left(\frac{\partial \Phi}{\partial e_2}\right)^2 + \frac{9}{4}\frac{1-\alpha^2}{e_2}\left(\frac{\partial \Phi}{\partial \alpha}\right)^2, \\
s_3 &= \frac{2}{\sqrt{3}}e_2^{3/2}\left\{\alpha\left(\frac{\partial \Phi}{\partial e_2}\right)^2 + \frac{9}{2}\frac{1-\alpha^2}{e_2}\left(\frac{\partial \Phi}{\partial e_2}\right)^2\frac{\partial \Phi}{\partial \alpha} - \right. \\
&\quad \left. -\frac{27}{4}\frac{\alpha(1-\alpha^2)}{e_2^2}\frac{\partial \Phi}{\partial e_2}\left(\frac{\partial \Phi}{\partial \alpha}\right)^2 - \frac{27}{8}\frac{1-\alpha^2}{e_2^3}\left(\frac{\partial \Phi}{\partial \alpha}\right)^3\right\}
\end{aligned}$$

or, in terms of explicitely introduced ψ,

$$s_1 = 3\frac{\partial \Phi}{\partial e_1},$$

$$s_2 = e_2\left\{\left(\frac{\partial\Phi}{\partial e_2}\right)^2 + \frac{1}{4e_2^2}\left(\frac{\partial\Phi}{\partial\psi}\right)^2\right\},$$

$$s_3 = \frac{2}{\sqrt{3}}e_2^{3/2}\left\{\left(\frac{\partial\Phi}{\partial e_2}\right)^2 \sin 3\psi + \frac{3}{2}\frac{1}{e_2}\left(\frac{\partial\Phi}{\partial e_2}\right)^2\left(\frac{\partial\Phi}{\partial\psi}\right)\cos 3\psi - \right. \tag{4.83}$$

$$\left. -\frac{3}{4}\frac{1}{e_2^2}\left(\frac{\partial\Phi}{\partial e_2}\right)\left(\frac{\partial\Phi}{\partial\psi}\right)^2 \sin 3\psi - \frac{1}{8}\frac{1}{e_2^3}\left(\frac{\partial\Phi}{\partial\psi}\right)^3 \cos 3\psi\right\}.$$

Denoting:

$$\chi = \frac{1}{2e_2}\frac{\partial\Phi}{\partial\psi}/\frac{\partial\Phi}{\partial e_2}, \tag{4.84}$$

the last two relations of (4.83) allow to construct $\sin 3\zeta$:

$$\sin 3\zeta = \frac{(1-3\chi^2)\sin 3\psi + (-\chi+3)\chi\cos 3\psi}{(1+\chi^2)^{3/2}}. \tag{4.85}$$

Choosing notation $\chi = \tan\omega$ in (4.83), some simple calculations yield:

$$\sin 3\zeta = \sin 3(\psi+\omega), \tag{4.86}$$

with the solution:

$$\omega = \zeta - 2\psi + \frac{2n\pi}{3},$$

where n is an integer. Taking $n = 0$, we have:

$$\omega = \zeta - 2\pi, \tag{4.87}$$

and relationship between the three quantities ω, ζ and ψ is now univocal. Introducing $\chi = \tan\omega$ into (4.83), these relations become:

$$s_1 = 3\frac{\partial\Phi}{\partial e_1},$$

$$s_2 = \frac{e_2}{\cos^2\omega}\left(\frac{\partial\Phi}{\partial e_2}\right)^2, \tag{4.88}$$

$$s_3 = \frac{2}{\sqrt{3}}e_2^{3/2}\frac{\sin 3(\psi+\omega)}{\cos^2\omega}\left(\frac{\partial\Phi}{\partial e_2}\right)^2.$$

It is useful for our final purpose to define the gravitational quantities:

$$K = \frac{1}{3}\frac{s_1}{e_1}, \quad G = \frac{1}{2}\sqrt{\frac{s_2}{e_2}}, \tag{4.89}$$

in which case formulas (4.88) become:

$$\begin{aligned}\frac{\partial\Phi}{\partial e_2} &= Ke_1,\\ \frac{\partial\Phi}{\partial e_2} &= 2G\cos\omega,\\ \frac{\partial\Phi}{\psi} &= 4Ge_2\sin\omega.\end{aligned}\tag{4.90}$$

With these observations, relations (4.81) write:

$$\begin{aligned}\frac{\partial\Phi}{\partial E_1} &= Ke_1 + \frac{4}{3}Ge_1\frac{\cos(3\psi+\omega)}{\cos 3\psi} - 2G\frac{e_2+\frac{1}{3}e_1^2}{\sqrt{3e_2}}\frac{\sin\omega}{\cos 3\psi},\\ \frac{\partial\Phi}{\partial E_2} &= -2G\left\{\frac{\cos(3\psi+\omega)}{\cos 3\psi} - \frac{e_1}{\sqrt{3e_2}}\frac{\sin\omega}{\cos 3\psi}\right\},\\ \frac{\partial\Phi}{\partial E_3} &= -2G\sqrt{\frac{3}{e_2}}\frac{\sin\omega}{\cos 3\psi}.\end{aligned}\tag{4.91}$$

In view of (4.91), relations (4.79) receive the following tensor form:

$$\widehat{D}_\sigma = 2G\left\{\frac{\cos(3\psi+\omega)}{\cos 3\psi}\widehat{D}_\varepsilon - \sqrt{\frac{3}{e_2}}\frac{\sin\omega}{\cos 3\psi}\left(\widehat{D}_\varepsilon^2 - \frac{2}{3}e_2\widehat{I}\right)\right\},\tag{4.92}$$

where $\widehat{D}$ denotes the deviator of the respective matrix, and $\widehat{I}$ the unit matrix. The matrix relation (4.92) stands for the most gravitational general constitutive relation for the gravitational ideal isotropic elastic material structure, but only for this one. Indeed, this relation does not concern only gravitational reversible transformations, since it has been obtained by means of the gravitational potential Φ, which can be settled not only for gravitational reversible transformations. It is also revertible: $\widehat{D}_\varepsilon$ can be expressed in terms of $\widehat{D}_\sigma$, as we shall further show. To this end, let us write $\widehat{D}_\varepsilon$ as a formal series in $\widehat{D}_\sigma$:

$$\widehat{D}_\varepsilon = \mu_0\widehat{I} + \mu_1\widehat{D}_\sigma + \mu_2\widehat{D}_\sigma^2 + \ldots\tag{4.93}$$

This series does not contain power of order higher than 3. Indeed, according to Hamilton-Cayley of algebra [6], any matrix is a solution of its characteristic equation. Applying this to $\widehat{D}_\sigma$ in terms of notations used above, we have:

$$\widehat{D}_\sigma^3 - s_2\widehat{D}_\sigma^2 + \frac{1}{3}s_3\widehat{I} = 0,\tag{4.94}$$

which says that any power than 3 of the deviator is expressed in terms of polynomials with at most second order in $\widehat{D}_\sigma$. According to (4.94):

$$\widehat{D}_\sigma^3 = s_2\widehat{D}_\sigma^2 - \frac{1}{3}s_3\widehat{I},$$

so that:

$$\widehat{D}_\sigma^4 = s_2\widehat{D}_\sigma^2 - \frac{1}{3}s_3\widehat{D}_\sigma,$$
$$\widehat{D}_\sigma^5 = s_2\widehat{D}_\sigma^3 - \frac{1}{3}s_2\widehat{D}_\sigma^2 = s_2^2\widehat{D}_\sigma - \frac{1}{2}s_2\widehat{D}_\sigma^2 - \frac{1}{2}s_2s_3\widehat{I}\text{...etc.}$$

Therefore, the most general form of (4.93) is:

$$\widehat{D}_\varepsilon = q_0\widehat{I} + q_1\widehat{D}_\sigma + q_2\widehat{D}_\sigma^2, \tag{4.95}$$

where q_0, q_1, q_2 are functions of the invariants of $\widehat{D}_\sigma$. In the last relation q_0 and q_2 cannot be independent, since the l.h.s. is a deviator, and, as a result, its trace has to be real. Whereas the eigenvalues of $\widehat{D}_\sigma$ are $2\sigma_1 - \sigma_2 - \sigma_3$, this demand requires:

$$q_0 = -\frac{2}{3}s_2q_2,$$

which means:

$$\widehat{D}_\varepsilon = q_1\widehat{D}_\sigma + q_2\left(\widehat{D}_\sigma^2 - \frac{2}{3}s_2\widehat{I}\right). \tag{4.96}$$

Extracting $\widehat{D}_\sigma$ from (4.92) and replacing into (4.96), we arrive at:

$$\widehat{D}_\varepsilon = \alpha\widehat{I} + \beta\widehat{D}_\varepsilon + \gamma\widehat{D}_\varepsilon^2$$

which has to be an identity, that is:

$$\alpha = \gamma = 0, \beta = 1,$$

so that the solution is:

$$q_1 = \frac{1}{2G}\frac{\cos(3\zeta - \omega)}{\cos 3\zeta}, \quad q_2 = \frac{1}{2G}\sqrt{\frac{3}{e_2}\frac{\sin\omega}{\cos 3\zeta}},$$

and (4.96) finally writes:

$$\widehat{D}_\varepsilon = \frac{1}{2G}\left\{\frac{\cos(3\zeta - \omega)}{\cos 3\zeta}\widehat{D}_\sigma + \sqrt{\frac{3}{s_2}\frac{\sin\omega}{\cos 3\zeta}}\left(\widehat{D}_\sigma^2 - \frac{2}{3}s_2\widehat{I}\right)\right\}. \tag{4.97}$$

For $\omega \equiv 0$, formula (4.92) yields the Hook's law of gravitational type for a perfectly isotropic material structure, by means of the identifications:

$$2G = \frac{E}{1+\mu}, \quad K = \frac{E}{1-2\mu},$$

where μ is gravitational type Poisson's coefficient:

$$\sigma_{ij} = E\left[\frac{1}{1+\mu}\varepsilon_{ij} + \frac{\mu}{(1+\mu)(1-2\mu)}(\varepsilon_{11}+\varepsilon_{22}+\varepsilon_{33})\right].$$

The irreversible plastic gravitational strain can be described by the following two hypotheses:

i) The plastic gravitational strain takes place without the modifying of the materials structure volume. If ε_i are the main gravitational strain, then this condition becomes:

$$\varepsilon_1 + \varepsilon_2 + \varepsilon_3 = 0. \tag{4.98}$$

ii) The maximum shearing gravitational strain are proportional with the maximum shearing gravitational stresses, *i.e.*:

$$\frac{\varepsilon_2 - \varepsilon_3}{\sigma_2 - \sigma_3} = \frac{\varepsilon_3 - \varepsilon_1}{\sigma_3 - \sigma_1} = \frac{\varepsilon_1 - \varepsilon_2}{\sigma_1 - \sigma_2} = c. \tag{4.99}$$

The constant c can depend both on the time and on the material's thermodynamic state parameters. It is assumed in Eqs. (4.99) that no change in the gravitational stresses and gravitational axes orientations takes place.

Only two equations from (4.99) are independent. Taking them into consideration along with Eq. (4.98) and solving the system for ε_i, we obtain:

$$\varepsilon_i = c \cdot s_i, \tag{4.100}$$

where:

$$s_i = \frac{1}{3}(2\sigma_i - \sigma_j - \sigma_k) \tag{4.101}$$

with (i, j, k) forming a circular permutation.

Now we can observe that relations (4.100), named gravitational constitutive relations, derive from a gravitational potential which has an gravitational energetic meaning. Indeed, it is considered that in a point in the material structure the gravitational stress tangent to a plane which has the normal given by the cos value for l_1, l_2, l_3 is:

$$\tau^2 = \sigma_1^2 l_1^4 + \sigma_2^2 l_2^4 + \sigma_3^2 l_3^4 - \left(\sigma_1 l_1^2 + \sigma_2 l_2^2 + \sigma_3 l_3^2\right)^2. \tag{4.102}$$

The spatial average of this value is defined by admitting a uniform distribution of the gliding planes normals:

$$\tau_m^2 = \frac{1}{\Omega}\int \tau^2 d\Omega, \tag{4.103}$$

where $d\Omega$ is the elementary quantity of the solid angle and can be expressed through:

$$d\Omega = \sin\theta d\theta d\varphi \tag{4.104}$$

and:

$$\begin{aligned} l_1 &= \sin\theta\cos\varphi, \\ l_2 &= \sin\theta\sin\varphi, \\ l_3 &= \cos\theta. \end{aligned} \tag{4.105}$$

By integrating the values from (4.103) we obtain [27]:

$$\tau_m^2 = \frac{1}{15}\left\{(\sigma_2 - \sigma_3)^2 + (\sigma_3 - \sigma_1)^2 + (\sigma_1 - \sigma_2)^2\right\}. \tag{4.106}$$

In a totally similar way it is shown that the average of the normal gravitational stress is given by:

$$\overline{p} = \frac{1}{3}(\sigma_1 + \sigma_2 + \sigma_3). \tag{4.107}$$

From relation (4.106) it can be shown that:

$$\frac{\partial \tau_m}{\partial \sigma_i} = \frac{1}{5\tau_m} \cdot s_i, \tag{4.108}$$

from which it results that for $\tau_m = \text{const.}$, it can be written that:

$$\varepsilon_i = k\frac{\partial \tau_m}{\partial \sigma_i}. \tag{4.109}$$

This means that ε_i is proportional with the "equipotential" surface gradient $\tau_m = \text{const.}$ from the gravitational stresses space. This surface is called material structure flow surface, and relation (4.109) is the flowing law associated to this surface.

It is a proven fact that the material structure gains a plastic flow around a value which, in one-axial traction, is called flowing gravitational stress $-\sigma_c$. Because neither the general stresses, or calculations, excepting overly special cases, nor the gravitational stresses induced by gravitational strains in a material structure are one-axial, it is necessary to find a three-axial equivalent for the one-axial gravitational strains. This equivalent naturally appears, if we take into account the previous observations, through τ_m, or, for more convenient reasons, through the quantity:

$$\bar{\sigma} = \frac{1}{\sqrt{2}} \left\{ \sum (\sigma_2 - \sigma_3)^2 \right\}^{\frac{1}{2}}. \tag{4.110}$$

Thus, if the following condition is satisfied:

$$\bar{\sigma} = \sigma_c, \tag{4.111}$$

the material structure gains a plastic flow. This condition constitutes

If, analog with relation (4.110), the gravitational strain, the von Mises criterium of gravitational type intensity is defined as:

$$\bar{\varepsilon} = \frac{\sqrt{2}}{3} \left\{ \sum (\varepsilon_2 - \varepsilon_3)^2 \right\}^{\frac{1}{2}} \tag{4.112}$$

then, from relation (4.99), it results:

$$c = \frac{3}{2} \frac{\bar{\varepsilon}}{\bar{\sigma}} \tag{4.113}$$

with which relation (4.99) takes the form:

$$\varepsilon_i = \frac{3}{2} \frac{\bar{\varepsilon}}{\bar{\sigma}} s_i. \tag{4.114}$$

This relation leads to an or theoretical search for the functional relationship:

$$\bar{\varepsilon} = f(\bar{\sigma}) \tag{4.115}$$

which can provide the coefficient from Eq. (4.114). Taking into account the above mentioned equivalency between the one-axial and three-axial cases, relation (4.115) can be usually considered as an gravitational curve or a one-axial gravitational compression.

The following problem arises: can there be an apriori theoretical relation between the one-axial gravitational strains case and the three-axial gravitational strains that can validate criterion (4.111)? In the following it will be shown that such a relation exists, from which it also naturally arises the validity limit for this criterion. In order to achieve this, the measurable continuous groups theory will be used, by applying the results of integral geometry [13].

The first issue in solving the above mentioned problem is the selection of a "standard" version of the gravitational curve $\varepsilon = f(\sigma)$. In order to achieve this we have selected the Bell solution of gravitational type, according to which, under the limit of big gravitational strain, the gravitational stress curve is the hull of a parabolas family of the form (it is our working hypothesis):

$$\sigma^2 = a^2(\varepsilon - b), \tag{4.116}$$

where a and b are constants that are depend for example, on the texture conditions of the material structure.

If we center around the parabolas family from Eq. (4.116), then it is clear that it must make its mark on a possible plane geometry (σ, ε). This geometry can be founded on a parametric group which must make the form from relation (4.116) invariant. This group can be best revealed if the homogenous coordinates $(\sigma, \varepsilon)$are used in the form:

$$\frac{x_1}{x} = \frac{x_2}{y} = \frac{x_3}{1}, \tag{4.117}$$

where:

$$\begin{aligned} x &= \varepsilon - b, \\ y &= \frac{\sigma}{a} \end{aligned} \tag{4.118}$$

case in which Eq. (4.116) becomes:

$$x_2^2 - x_1 x_3 = 0. \tag{4.119}$$

In this situation, the conic from relation (4.119) accepts the canonic parameterization:

$$\frac{x_1}{t^2} = \frac{x_2}{t} = \frac{x_3}{1}, \tag{4.120}$$

where t is a real parameter, and its invariance group is the three-parameters group generated by the homographic transformation of the t parameter. If this transformation is written under a more convenient form:

$$t = \frac{t + \alpha_1}{1 - \alpha_2 - \alpha_3}, \tag{4.121}$$

which highlights the unit transformation for $\alpha_1 = \alpha_2 = \alpha_3 = 0$, then using Eq. (4.120) the following transformation relations for the parameters x_1, x_2 result:

$$x_1 = \frac{x_1 + 2\alpha_1 x_2 + \alpha_1^2}{\alpha_3^2 x_1 - 2\alpha_3 (1-\alpha_2) x_2 + (1-\alpha_2)^2},$$
$$x_2 = \frac{-\alpha_3 x_1 + (1-\alpha_2-\alpha_1\alpha_3) x_2 + \alpha_1 (1-\alpha_2)}{\alpha_3^2 x_1 - 2\alpha_3 (1-\alpha_2) x_2 + (1-\alpha_2)^2}, \tag{4.122}$$

from which a continuous two variables with three parameters group can be observed. The Lie algebra [13,27] is given by the operators:

$$\widehat{L}_1 = 2y\frac{\partial}{\partial x} + \frac{\partial}{\partial y},$$
$$\widehat{L}_2 = 2x\frac{\partial}{\partial x} + y\frac{\partial}{\partial y}, \tag{4.123}$$
$$\widehat{L}_3 = 2xy\frac{\partial}{\partial x} + \left(2y^2 - x\right)\frac{\partial}{\partial y},$$

with the commutation relations:

$$[\widehat{L}_1, \widehat{L}_2] = \widehat{L}_1,$$
$$[\widehat{L}_2, \widehat{L}_3] = \widehat{L}_3, \tag{4.124}$$
$$[\widehat{L}_3, \widehat{L}_1] = -2\widehat{L}_2,$$

where inhomogeneous coordinates were taken into account in order to simplify the writing.

As it should be, the conics in relation (4.119) appear in this situation as the Eq. (4.123) group's invariant varieties with two parameters, and this is why they are invariant only with regard to the first two operators from relation (4.123). The issue at hand is not to find the two-parameters invariant varieties families, but to find the three-axial that holds three parameters: the main gravitational stresses, *i.e.* the eigenvalues of the gravitational stress tensor. Now the gravitational stresses evolution group remains to be solved, which must be isomorphic to the group from Eq. (4.123). In order to highlight it we must note that the main gravitational stresses are the solution to the secular equation of the respective matrix, which can be written as:

$$\sigma^3 + 3a_1\sigma^2 + 3a_2\sigma + a_3 = 0, \tag{4.125}$$

where $3a_1, 3a_2, a_3$ are the orthogonal invariants of the gravitational stresses matrix. If the gravitational stresses state varies from $\sigma_1, \sigma_2, \sigma_3$

to $\sigma_1', \sigma_2', \sigma_3'$ then an algebra theorem [2], [3], [27] shows that between the secular equations, which have the respective values as roots, a linear relation takes place, generated by the homographic transformation:

$$\sigma' = \frac{a\sigma + b}{c\sigma + d}, \tag{4.126}$$

which gives a three-parameters group but in three variables. By writing the roots of the curve from relation (4.125) in the Barbilian form [27]:

$$\sigma' = \frac{h + \varepsilon_i \bar{h} k}{1 + \varepsilon_i k}, \tag{4.127}$$

where $\varepsilon_i^3 = 1$, $h, \bar{h}$ are quantities conjugated one to the other, and k is a one-module complex factor, the transformation from Eq. (4.126) induces upon the quantities $h, \bar{h}, k$ the real transformations (4.51), *i.e.* the Barbilian group.

This group is simple transitive, with the infinitesimal generators given by the operators (3.85) which reveals for the associated Lie algebra a structure that is identical with the one from Eq. (4.124). Therefore the two groups are isomorphic, the operators (4.123) and (3.85) being generated by the one and the same algebra (4.119). Moreover, the group (3.85), being simple transitive, is definitely measurable, its elementary measure being given by the differential three-form (4.52).

The issue now at hand is to find the invariant varieties families of the group (4.123) with three parameters, having group (3.85) associated as a parameters group. In our opinion these functions can provide for an answer to the problem of the correlation between the one-axial behavior of the gravitational stresses-strains curve and the gravitational stresses induced in the material structure by the one-axial gravitational strain.

These varieties families will be solutions of the Stoka [27] equations:

$$\begin{aligned}
&2y\frac{\partial f}{\partial x} + \frac{\partial f}{\partial y} + \frac{\partial f}{\partial h} + \frac{\partial f}{\partial \bar{h}} = 0,\\
&2x\frac{\partial f}{\partial x} + y\frac{\partial f}{\partial y} + h\frac{\partial f}{\partial h} + \bar{h}\frac{\partial f}{\partial \bar{h}} = 0,\\
&2xy\frac{\partial f}{\partial x} + \left(2y^2 - x\right)\frac{\partial f}{\partial y} + h^2\frac{\partial f}{\partial h} + \bar{h}^2\frac{\partial f}{\partial \bar{h}} + \left(h - \bar{h}\right)k\frac{\partial f}{\partial k} = 0.
\end{aligned} \tag{4.128}$$

This system admits solutions of the form:

$$f\left(\alpha, k_0^2\right) = \text{const.} \tag{4.129}$$

where:

$$\alpha = \frac{\sqrt{x - y^2}\left(h - \bar{h}\right)}{x - \left(h - \bar{h}\right)y + h\bar{h}},$$
$$k_0^2 = k^2 \frac{x - 2y\bar{h} + \bar{h}^2}{x - 2yh + h^2}. \tag{4.130}$$

It can be observed that the last of these integrals is a one-module complex one. In principle, f can be any function which is continuous and derivable in its variables. It is not yet known what kind of interpretation can a general solution such as Eq. (4.129) can have, but some specific integrals values from relation (4.130) can still be interpreted. Thus, if the one-axial gravitational strain is monotonous, then Eq. (4.130) must fulfill the condition $y^2 = x$, fact which leads to the specific value $x = 0$. In this case, the second relation (4.130) gives:

$$k_0 = k \frac{y - \bar{h}}{y - h}, \tag{4.131}$$

from which we can write y as:

$$y = \frac{\bar{h}k - hk_0}{k - k_0}. \tag{4.132}$$

The result we obtained in this case is important mainly because it shows that y can be identified in a specific case with one of the main gravitational stresses values. Indeed, if $k_0 \equiv (-1, -\varepsilon, -\varepsilon^2)$ then the situation from Eq. (4.127) is again reached. Therefore we can state that in both these specific cases the gravitational stress in the one-axial gravitational strain can be considered as one of the internal gravitational stresses eigenvalues. However we can draw more from Eq. (4.132). If this equation is written for $k_0 = -1$:

$$y = \frac{h + \bar{h}k}{1 + k}, \tag{4.133}$$

and $h, \bar{h}, k$ are explicitly written with regard to the main gravitational stress, and also the system of equations (4.127) is solved with regard to $h, \bar{h}, k$, then the following relations can be found:

$$h = -\frac{\sigma_2\sigma_3 + \varepsilon\sigma_3\sigma_1 + \varepsilon^2\sigma_1\sigma_2}{\sigma_1 + \varepsilon\sigma_2 + \varepsilon^2\sigma_3},$$
$$k = \frac{\sigma_1 + \varepsilon^2\sigma_2 + \varepsilon\sigma_3}{\sigma_1 + \varepsilon\sigma_2 + \varepsilon^2\sigma_3}. \tag{4.134}$$

These can be related with the above mentioned parameters by:

$$k = -e^{2i\xi},$$
$$h = \overline{p} + 2i\sigma\cos\xi, \tag{4.135}$$

where:

$$\tan\xi = \frac{2\sigma_1 - \sigma_2 - \sigma_3}{\sqrt{3}\,(\sigma_2 - \sigma_3)}, \tag{4.136}$$

and p, σ are given in relations (4.107) and (4.110). The quantity from Eq. (4.136) is the known Lode Nadai parameter of gravitational type.

If we use Eqs. (4.135) in relation (4.133), we obtain:

$$y = \overline{p} + \sigma\sin\xi \tag{4.137}$$

relation which, in the case of the absence of hydrostatic gravitational strain, is reduced to:

$$y = \sigma\sin\xi. \tag{4.138}$$

From this it result that the gravitational stress found under a one-axial monotonous gravitational strain can be identified with the quantity σ from Eq. (4.110) only if the gravitational strains which are orthogonal to the direction of gravitational stress are very close to each other. Indeed, in this case, from relation (4.136) it results that $\tan\xi \to \infty$, and, thus, $\sin\xi \to 1$. Therefore the von Mises criterion of gravitational type is conditioned by the previous restrictions. Hence the answer to the question posed above is a positive one: the one-axial tension can be identified with the quantity σ.

In such perspective it results:

i) The von Mises criterion of gravitational type is important in more than one aspects: the special parabolas in gravitational strain and the gravitational strain curves specificity – Bell's relation of gravitational type. It is to be expected that this criterion will not be satisfied in the case of gravitational compression;

ii) Limiting only to Bell's relation of gravitational type as being universal in the theory of space-time, the von Mises criterion of gravitational type is still important, by the selection of k_0. By employing this quantity the initial state in gravitational deformation can, for example, be introduced, which might justify a Hill or a more general type criterion;

iii) In the case of cyclic gravitational strains, which are outside the monotonous gravitational strain curve, certainly the α parameter is non-null, and therefore, it is of great importance for us to find a functional relation between it and k_0^2. Certainly the gravitational constitutive laws of this case greatly differ from the ones in the monotonous gravitational strains, and the method proposed above opens up a path for obtaining a gravitational constitutive law.

iv) Von Mises criterion of gravitational type becomes also a separation criterion of the geometrical frames from physical one, and finally, a separation criterion of space from matter.

References

1. Agop, M., Mazilu, N.: C.R.Sc. Paris, t. 304, Série II, nr. 9, 395, (1987).
2. Barbilian, D.: Opera Didactica, Technical Publishing House, Bucharest, Vol. I, II, III, (1968), (1971), (1974).
3. Barbilian, D.: Mathematica, Romanian Academy Publishing House, Bucharest, vol. I, (1967).
4. Beju, I., Sóos, E., Teodorescu P. P.: Techniques of Euclidean tensor calculus with applications (in Romanian), Technical Publishing House, Bucharest, (1977).
5. Carruthers, P., Nieto, M. M.: Rev. Mod. Phys., 40, 411, (1968).
6. Casanova, G.: L'Algèebre Vectorielle, Mir, Moscou, (1979).
7. De Broglie, L.: La Thermodinamique de la particule isolée, Gauthier-Villars, Paris, (1966).
8. Dumitru, S.: Microphysics, Dacia Publishing House, Cluj- Napoca, (1984) (in Romanian).
9. Ernst, F. J.: J. Math. Phys., 12, 239, (1971).

10. Gogala, B.: Int. J. Theor, Phys., 19, 573, (1980).

11. Ionescu-Pallas, N.: General Relativity and Cosmology, Scientific and Encyclopedic Publishing House, Bucharest, (1980) (in Romanian).

12. Jaynes, E. T.: Found. Phys., 3, 477, (1973).

13. Mihăileanu, N.: Analytical, differential and projective geometry. Complements, Didactic and Pedagogical Publishing House, Bucharest, (1972).

14. Mazilu, N.: Internal Report I.R.N.E., Piteşti, (1983) (in Romanian).

15. Rindler, W.: Amer. J. Phys., 34, 1174, (1966).

16. Schwartz, H. M.: Int. J. Theor, Phys., 16, 249, (1977).

17. Sobczyk, G.: Acta Phys. Pol., B, 12, 509, (1984).

18. Stoler, D.: Phys. Rev., D, 1, 3217, (1970).

19. Vrânceanu, G., Lectures on differential geometry, Didactic and Pedagogical Publishing House, Bucharest, (1979), Vol. III (in Romanian).

20. Yamamoto, T.: Progr. Theor. Phys., (Japan), 8, 258, (1952).

21. Wilhelm, H. E.: Phys. Rev., D1, 2278, (1970).

22. Stoler, D.: Phys. Rev., D4, 1971, 1925.

23. Gottlieb, I., Agop, M., Buzdugan, M., Crăciun, P.: Chaos, Solitons and Fractals, 24, 391, (2005).

24. Agop, M., Vasilica, M.: El Naschie's supergravity by means of the gravitational instantons synchronization. Chaos Solitons and Fractals, 30, 318, (2006).

25. Agop, M., Rusu, I.: Chaos, Solitons and Fractals, 34, 172, (2007).

26. Agop, M., Abacioaie, D.: El Naschie's space-time, interface between Weyl-Dirac bubbles and fractal superstrong, Cahos Solitons and Fractals, 34, 235, (2007).

27. Mazilu, N., Agop, M.: Skyrmions. A Great Finishing Touch to Classical Newtonian Philosophy, New York: Nova Science Publishers, (2012).

28. Mercheş, I., Agop, M.: Differentiability and Fractality in Dynamics of Physical Systems, World Scientific, Singapore, (2015).

29. Agop, M., Mazilu, N.: Fundamente ale Fizicii Moderne, Editura Junimea, Iaşi, (1989).

30. Novozhilov, V. V.: Theory of Elasticity, Pergamon Press and Macmillan, New York, (1961).

CHAPTER 5

Role of surface gauging in extended particle interactions: the case for spin

5.1 Mater and space

The quality of the matter first striking our senses is its space extent: every physical body we notice in our environment occupies a finite space. From a theoretical point of view one cannot talk of the finiteness of a body, without accepting, at least implicitly, a surface delimiting it. Even though, for a long time now, the theoretical physicists have understood the crucial role of this surface; with the present work we push this understanding to its extreme. Specifically, we document here the statement that not only the surface has the essential role of demarcation of a finite body, but when it comes to the description of a constituent particle of a material structure, it plays a fundamental theoretical part, which we have to consider more closely. Mention should be made, however, that the idea of surface here is not to be thought exclusively in plain geometrical terms.

There are indeed, in this respect, two special instances where we can make the best case for physics. First of all we have the separation of matter from the empty space: this is by no means a neatly defined surface, but what we propose to call a limit of separation. It is actually a notorious place of wild inhomogeneity, whereby the separation was occasionaly thought of even as a layer of finite thickness, to be defined by a variable density of matter. Poisson was the first to point out the necessity of the theoretical account of material phenomena characterized by the variable density of matter [1]. From the point of view of the existing differential theories of surfaces this situation can be pictured as a structural "evolution" of a geometric surface along a certain transversal direction in space, not necessarily normal, or even along a certain transversal curve in space, over a finite distance. This evolution, however, makes the concept of normal to surface, and even that of general orthonormal frame for that matter, obsolete: the evolution of a surface cannot take place strictly along the Euclidean normal to a point. In other words, from a geometrical point of view, a physical limit of separation presents, in a certain point, different structural properties in different directions in space.

The second instance serving the case for physics in the differential theory of surfaces is that of particle interactions. In any interaction, we have, first and foremost the idea of target particle, like, for example, a target nucleus. This particle sets a local reference frame for the surrounding space, first of all by an ideal fixed point the origin with respect to which we reckon the directions in that space. Certainly, then, from a physical point of view, it makes sense to assume that a nucleus has a space extent, and therefore a limit of separation from the empty space, which can be thought of as a "dynamic layer" in the manner described above. A projectile moving towards the target "sees" a succession of geometrical structures in that layer, and the close physical interaction can be described as a "layer interaction" so to speak, inasmuch as the projectile itself is to be conceived as an extended particle, just like the target itself.

We have done some theoretical studies lately (see [2]), and the references therein), which allow us to hope for a "holographic" theory that closely parallels the philosophical point of view of the ever-changing structure of the matter. The general idea of such a theory would be that a material structure is made of extended particles. Both the particles themselves and the structure they determine are ephemeral, inasmuch as they change continuously. The experience shows that the changes of a particle are reflected in its limit of separation from space. It is therefore only natural to think that the main theoretical tool of theoretical description of the matter should be related to this "skin" of the structural particles. The present chapter describes the meaning of these statements in their essentials.

5.2 Essentials of a natural embedding of surfaces in space

In classical differential theory of the surface geometry, one usually takes three mutually orthogonal unit vectors $\widehat{\mathbf{e}}_1, \widehat{\mathbf{e}}_2$ and $\widehat{\mathbf{e}}_3$ as a reference frame in space. Then, in order to describe a surface locally - to imbed it, as they say - we have to adapt such a reference frame to it, by appropriate rotations and translations. The classical differential geometry of surfaces always uses the notion of metric, as it usually works in the Euclidean space, and thus the definition of an orthonormal frame of reference is always at hand. So, for the moment we undertake a brief review of the local differential geometry of surfaces from such a standpoint, just in order to clarify our own point of view. The best gradual description of the geometrical procedure to constructing such a theory can be found in the extended work of Michael Spivak on differential geometry, especially in the Chapter 2 of the Volume III, pp. 75ff [3].

Assume that $(\widehat{\mathbf{e}}_1, \widehat{\mathbf{e}}_2)$ is an orthonormal frame on a surface imbedded in the usual threedimensional space, whose unit mormal is $\widehat{\mathbf{e}}_3$. Usually, each displacement vector can be written in the form

$$d\mathbf{r} = s^k \widehat{\mathbf{e}}_k \equiv s^\alpha \widehat{\mathbf{e}}_\alpha + s^3 \widehat{\mathbf{e}}_3, \tag{5.1}$$

where $\mathbf{r}$ is the position vector of a generic point of surface. We adopt here the convention that the Latin indices run through values 1, 2, 3, thus referring to the ambient space, while the Greek indices refer to surface, and run only through values 1 and 2. We also use the summation convention over repeated indices all throughout this work. The last identity in equation (5.1) allows one to define a connection of the position in space with the surface. Specifically, the natural assumption of the Euclidean differential geometry of surfaces is that in order that the position vector remains on the surface, its component normal to surface should always be zero, so that we can write

$$d\mathbf{r} = s^\alpha \widehat{\mathbf{e}}_\alpha; \;\; s^3 = 0. \tag{5.2}$$

The relation of the surface with its environment is then recognized by the normal displacement of a certain point of the surface. This normal displacement is in turn described by the second fundamental form, according to the Cartan's procedure, which can be crudely described as follows.

The equation of evolution of this orthonormal frame can be written as

$$|d\widehat{\mathbf{e}}\rangle = \boldsymbol{\Omega}\cdot|\widehat{\mathbf{e}}\rangle;\quad \boldsymbol{\Omega}+\boldsymbol{\Omega}^t = 0, \tag{5.3}$$

where $\boldsymbol{\Omega}$ is a matrix of differential 1-forms, and the superscript "t" stands for "transposed". The last equality here is a direct consequence of the orthonormality of the frame. Now, considering equation (5.1) as an immersion equation, its conditions of integrability come down to a system of exterior differential equations, which can be written as

$$d\wedge s^\mu + \omega^\mu_\nu \wedge s^\nu = 0 \;\therefore\; \begin{array}{l} d\wedge s^1 + \omega^1_2 \wedge s^2 = 0,\\ d\wedge s^2 + \omega^1_2 \wedge s^1 = 0,\\ s^1\wedge \omega^3_1 + s^2\wedge\omega^2_3 = 0, \end{array} \tag{5.4}$$

where ω^β_μ are the entries of matrix $\boldsymbol{\Omega}$. Here the last conditions from equations (5.2) and (5.3) were used. The bottom equality from (5.4) leads, via the Cartan's Lemma, to the conclusion that there exists a convenient symmetric matrix h, such that

$$\omega^3_i = h_{ik}s^k;\quad h_{ij} = h_{ji}. \tag{5.5}$$

On the other hand, the skew symmetry of the matrix $\boldsymbol{\Omega}$ shows that this equation practically represents the variation of the local normal vector to surface, i.e. what we would like to call the curvature vector. Indeed, from equation (5.3) we have for the variation of this vector

$$d\widehat{\mathbf{e}}_3 = \omega^1_3\widehat{\mathbf{e}}_1 + \omega^2_3\widehat{\mathbf{e}}_2,\quad \omega^\alpha_3 + \omega^3_\alpha = 0. \tag{5.6}$$

Then differentiating once more the first equality from equation (5.2) we get

$$d^2\mathbf{r} = (ds^k + s^\alpha\omega^\beta_\alpha)\widehat{\mathbf{e}}_\beta + (s^\alpha\omega^3_\alpha)\widehat{\mathbf{e}}_3 \;\therefore\; \widehat{\mathbf{e}}_3\cdot d^2r = s^\alpha\omega^3_\alpha = h_{\alpha\beta}s^\alpha s^\beta. \tag{5.7}$$

The last expression here is the usual second fundamental form. It is the component of the differential of displacement (the second differential of the position vector in space) along the normal to surface.

This Cartanian approach of the differential geometry of surfaces has a tremendous advantage from theoretical physics' point of view. This advantage is offered here by the fact that the matrix $\mathbf{h}$ can be anything convenient. True, when one comes to evaluating it, one uses purely geometrical methods, but the geometry here is just a means of measurement, so to speak. One can consult [4] for an instructive example of evaluation of the curvature matrix from geometrical point of view. The very determination of the matrix **h** should nevertheless be a matter of physics. Our standpoint is that the three entries of the matrix **h** - here the curvature parameters - are externally defined quantities, allowing us to introduce physics, no matter if we usually evaluate them by geometrical means. As the structure of matter is always in evolution for some physical reasons, such an evolution should be first recognized in changes of its delimiting surface, viz. the surface of the constituent particles of its momentary structure.

Theoretically, one can imagine the evolution of a material structure as having three components. First, there are two components of the evolution proper: the evolution of the matter inside the constituent particles and the evolution of the structure to which they belong as constituent particles. Then, there is a relational component between the two kinds of evolution, in order to accommodate with each other. The theoretical description of this last component involves the limit of the constituent particles, therefore, at least ideally speaking, a surface, and a closed one for that matter, inasmuch as a particle is finite in every direction in space. Which brings us to the conclusion that the theoretical physics of the evolution of matter should be described in terms of a differential geometry of surfaces, more precisely by a geometry of the curvature parameters' space. In view of the brief presentation right above, we would like therefore a geometrical theory where not only the curvature, but also the surface metric is open to physical determination. This means extending the very differential theory of surfaces, a topic that the Cartanian approach of the theory of surfaces has brilliantly accomplished through the affine differential theory.

5.3 An affine differential generalization

The affine differential theory of surfaces is suggested by the very idea of reference frame, and the fixed point serving as its origin, which can also serve as a center, for instance as a center of force. In an affine theory we miss first of all a metric, which needs to be defined by external means. One can guess that these external means amount, naturally, to a physics of the problem at hand. Once we have a metric, we can define the relative directions, and therefore a normal to surface. In keeping with our manner of introducing the physical considerations into geometrical theory, we shall follow here a purely exterior differential form development - a Cartanian approach, as we call it here - available in quite a few remarkable works descending from a seminal one of S. S. Chern (see [5]). In broad lines, this philosophy can be presented as follows, starting from the idea of volume of a reference frame.

As a reference frame one chooses three vectors in space, with the only requirement of being linearly independent, and delimiting a certain volume theoretically represented as their mixed product. A reference frame is usually further defined by the essential condition that this volume should be constant either for a specific frame, or even for a family of frames. Further on, one maintains the definition of surface as in equation (5.2) above. However, because the frame is no more orthonormal, when one assumes its evolution like in the equation (5.3), the matrix $\mathbf{\Omega}$ is no more skew symmetric. The condition of definition of the frame by maintaining its constant volume gives instead the constraint:

$$tr(\mathbf{\Omega}) \equiv \omega_i^i = 0. \tag{5.8}$$

The structural equations are, as usual, the integrability conditions for the definition of differential of position and for the evolution of frame. These amount to

$$d \wedge s^j + \omega_k^j \wedge s^k = 0, \quad d \wedge \omega_i^j + \omega_k^j \wedge \omega_i^k = 0. \tag{5.9}$$

Now, assuming that our surface is convex (for a clear explanation of this concept see [3]), the last condition from equation (5.2) leads again to equation (5.5). However, this equation is here taken as defining a metric rather than a second fundamental form as in the regular Euclidean theory of surfaces. This metric is affinely invariant if we write it in the form [5,6]

$$(ds)^2 \equiv h^{-\frac{1}{4}} \omega_\alpha^3 s^\alpha = h^{-\frac{1}{4}} h_{\alpha\beta} s^\alpha s^\beta, \tag{5.10}$$

where h is the determinant of $\mathbf{h}$. One can see indeed, by comparison with equation (5.7) from the Euclidean differential theory of surfaces sketched above, that this quadratic form would be actually the equivalent of the second fundamental form of a regular surface. This is why, in the affine theory of surfaces, it is usually designated with II among geometers (see [5]).

The problem now remains to deal with the other side, namely ω_3^α, of the matrix $\mathbf{\Omega}$, because now this matrix is no more skew symmetric

$(\omega_3^\alpha + \omega_\alpha^3 \neq 0)$. From equation (5.6) above, one can see that these entries of the matrix $\mathbf{\Omega}$ are actually the proper counterparts of the components of the normal from the case of an Euclidean surface. One approach to solving this issue is to choose o_3^3 an exact differential, which means $d \wedge \omega_3^3 = 0$, in which case, from the corresponding equation (5.10), we have

$$\omega_3^1 \wedge \omega_1^3 + \omega_3^2 \wedge \omega_2^3 = 0 \therefore \begin{pmatrix} \omega_3^1 \\ \omega_3^2 \end{pmatrix} = \begin{pmatrix} b^{11} & b^{12} \\ b^{12} & b^2 \end{pmatrix} \begin{pmatrix} \omega_1^3 \\ \omega_2^2 \end{pmatrix}. \tag{5.11}$$

Here Cartan's Lemma, guaranteeing the existence of a convenient matrix b, was used again. This leads to the definition of a correspondent of the third fundamental form from the regular theory of surfaces, i.e. the square of what we have called earlier the curvature vector [7]. Using here equation (5.5), we can get from (5.11) this quadratic form:

$$\omega_3^i = b^{ij} h_{jk} s^k \equiv b_k^i s^k \therefore \mathrm{III} \equiv \omega_3^k \omega_k^3 = b_{ij} s^i s^j, \tag{5.12}$$

where the notation III seems to be, again, geometers' preference.

Now, having a metric, we can define an orthonormal frame in the tangent plane of the surface, $(\widehat{\mathbf{e}}_1, \widehat{\mathbf{e}}_2)$ say, and attach a space vector n transversal to surface in a general direction, such that the volume of $(\widehat{\mathbf{e}}_1, \widehat{\mathbf{e}}_2, \mathbf{n})$ is nonzero and remains constant. Then, the following property can be proved [6]: there is a unique space vector n and some adequate differential 1-forms ϕ^α and ψ^α, satisfying the structural equations

$$\mathbf{dm} = \phi^1 \widehat{\mathbf{e}}_1 + \phi^2 \widehat{\mathbf{e}}_2; \quad \begin{aligned} d\widehat{\mathbf{e}}_1 &= \omega_1^1 \widehat{\mathbf{e}}_1 + \omega_1^2 \widehat{\mathbf{e}}_2 + \phi^1 \mathbf{n}, \\ d\widehat{\mathbf{e}}_2 &= \omega_2^1 \widehat{\mathbf{e}}_1 + \omega_2^2 \widehat{\mathbf{e}}_2 + \phi^2 \mathbf{n}, \omega_1^1 + \omega_2^2 = 0. \end{aligned} \tag{5.13}$$

One can see from this system that the vector $\mathbf{n}$ is, formally at least, as close as possible to the normal vector from the regular differential theory of surfaces - the curvature vector dn is an intrinsic vector - and it is indeed geometrically known as the affine normal to surface. Given any affine frame, $(\mathbf{e}_{10}, \mathbf{e}_{20}, \mathbf{e}_{30})$ say, the affine normal is uniquely defined by equation

$$\mathbf{n} = a^1 \mathbf{e}_{10} + a^2 \mathbf{e}_{20} + h^{\frac{1}{4}} \mathbf{e}_{30}, \tag{5.14}$$

where the auxiliary vector $|a\rangle$ serving for this definition is determined by the differential equation:

$$\langle a|\mathbf{h}|\phi\rangle + d(h^{\frac{1}{4}}) + h^{\frac{1}{4}} o_3^3 = 0. \tag{5.15}$$

We can see here that, given the metric, one can either define $|a\rangle$ by choosing ω_3^3, or ω_3^3 by choosing $|a\rangle$. The geometers' preference seems to be the first procedure [8]. Indeed, equation (5.15) can be rewritten in the form

$$\langle a|\mathbf{h}|\phi\rangle + h^{\frac{1}{4}}\{\frac{1}{4}d\ln(h) + \omega_3^3\} = 0$$

and if one chooses the exact differential ω_3^3 such that

$$\frac{1}{4}d\ln(h) + \omega_3^3 = 0, \tag{5.16}$$

then an affine invariant normal vector can be defined with one further special choice, namely $|a\rangle = |0\rangle$, as:

$$\mathbf{n} = h^{\frac{1}{4}}\mathbf{e}_3. \tag{5.17}$$

This choice is particularly attractive in geometrical exploits of this mathematical theory, for in cases where $h = 1$, we can write

$$\omega_3^3 = 0, \quad \mathbf{n} = \widehat{\mathbf{e}}_3. \tag{5.18}$$

Notice, however that, while still under the condition of affine invariance, but with $h \neq$ constant, the equation (5.18) can offer ω_3^3 as the differential of a function depending exclusively on some external parameters already introduced through the metric tensor. Therefore the choice of ω_3^3 as an exact differential can be subjected to the very same physical considerations to which the matrix $\mathbf{h}$ is subjected.

Moreover, even if we give up the affine invariance for the definition of the affine normal, we still have other possibilities of defining the vector $|a\rangle$, depending on the location in the tangent plane of the affine surface. For once, these posssibilities can be offered, for instance, by the nontrivial solutions of the equation

$$\langle a|\mathbf{h}|\phi\rangle = 0 \quad \frac{a^1}{h_{12}\phi^1 + h_{22}\phi^2} = \frac{-a^2}{h_{11}\phi^1 + h_{12}\phi^2}. \tag{5.19}$$

However, there are still other possibilities, to be exploited here, upon which we do not insist momentarily. The bottom line is that the affine theory of surfaces has this attracting feature, excellently serving our physical point of view that, by a Cartanian approach, both the metric and the curvature are totally external things, so that they can both carry a physical content.

5.4 A definition of the metric tensor: the Fubini-Pick cubic differential form

One important concept that we were set out to exploit in a previous work [7] is that of algebraic apolarity. The affine differential theory of surfaces makes obvious that the apolarity involves a cubic differential invariant - the so-called Fubini-Pick cubic form. This differential form can be best introduced in the Cartanian manner that follows [5]. By exterior differentiating equation (5.8), within the choice from equation (5.13), and using the corresponding equality from equation (5.5), we get

$$\mathrm{D}h_{\alpha\beta} \wedge \phi^{\beta} = 0, \quad \mathrm{D}h_{\alpha\beta} \equiv dh_{\alpha\beta} - h_{\alpha\mu}\omega_{\beta}^{\mu} - \omega_{\alpha}^{\nu}h_{\nu\beta}. \tag{5.20}$$

Using again Cartan's Lemma we get, for a convenient three-index tensor $\mathbf{\Phi}$:

$$\mathrm{D}h_{\alpha\beta} = \Phi_{\alpha\beta\gamma}\phi^{\gamma}. \tag{5.21}$$

The tensor $\mathbf{\Phi}$ is obviously symmetric in all three indices, as required by its definition. Geometrically, it measures the difference between the affine connection of the surface and the Levi-Civita connection, $\boldsymbol{\omega}$ say of the metric h, for which $\mathbf{Dh} = 0$:

$$\overline{\omega}_{\alpha}^{\beta} = \omega_{\alpha}^{\beta} + \frac{1}{2}h^{\beta\nu}\Phi_{\alpha\nu\sigma}\phi^{\sigma}. \tag{5.22}$$

The cubic differential form

$$\mathbf{\Phi} \equiv \Phi_{\alpha\nu\sigma}\phi^{\alpha}\phi^{\nu}\phi^{\sigma} \tag{5.23}$$

is known as the Fubini-Pick form of surface. Another important property of this form is that if it vanishes identically for a certain affine surface, then that surface is a quadric [8]. We think that it is mostly this property that came in handy in the past physical problems, as we shall see a little later.

However, by far the most important property of the Fubini-Pick form is, as we have already said, the property of apolarity, which leads directly to a physical way of defining the metric tensor. The procedure can be described as follows: if we take $\omega_3^3 = 0$, then by equation (5.16) we have $d\ln(h) = 0$, which means $tr(\mathbf{h}^{-1}d\mathbf{h}) = 0$. This is to say that the metric of an affine surface is apolar to the quadratic form having as coefficients the variations of the entries of the metric tensor. Now, using from equation (5.13) the fact that the affine connection of surface is trace-free, and the definition (5.21) of the Fubini-Pick tensor, the apolarity condition comes down to

$$h^{\mu\nu}\Phi_{\mu\nu\lambda} = 0. \tag{5.24}$$

This matrix equation can be considered an algebraic homogeneous system for the entries of the metric tensor:

$$h_{22}a_0 - 2h_{12}a_1 + h_{11}a_2 = 0, \quad h_{22}a_1 - 2h_{12}a_2 + h_{11}a_3 = 0. \tag{5.25}$$

Here we have used the definition of the inverse of the metric tensor, its property of symmetry, and the symmetries of the Fubini-Pick tensor, with the following notations:

$$\begin{aligned}\Phi_{111} &\equiv a_0, \quad \Phi_{112} = \Phi_{121} = \Phi_{211} \equiv a_1, \\ \Phi_{221} &= \Phi_{122} = \Phi_{212} \equiv a_2, \quad \Phi_{222} \equiv a_3.\end{aligned} \tag{5.26}$$

With these notations the Fubini-Pick cubic can be written in the form

$$\Phi \equiv a_0(\Phi^1)^3 + 3a_1(\Phi^1)^2\Phi^2 + 3a_2\Phi^1(\Phi^2)^2 + a_3(\Phi^2)^3. \tag{5.27}$$

If we assume that this cubic is known, which means that its coefficients are known, then from equation (5.25) we can calculate the affine metric up to an arbitrary factor:

$$\frac{h_{11}}{a_0a_2 - a_1^2} = \frac{2h_{12}}{a_0a_3 - a_1a_2} = \frac{h_{22}}{a_1a_3 - a_2^2}. \tag{5.28}$$

This purely algebraic result gives the entries of the affine metric tensor as coefficients of the Hessian of Fubini-Pick cubic [9], and shows that the affine metric of surface is actually the Hessian of Fubini-Pick cubic form. Summarizing, the apolarity in the case of affine surfaces comes down to the following statement: for the case of null 33 component of the matrix $\mathbf{\Omega}$, and constant surface element (h = constant), the affine metric form is the Hessian of Fubini-Pick cubic.

Obviously, this result can be used in the theory, in order to determine the metric of surface. Indeed, if by a physical reason we have at our disposal a cubic equation for the description of the matter, then this equation can be taken as a Fubini-Pick form in developing a metric theory of the surface delimiting the structure of that matter. The physics has a host of cases, mostly in the mechanics of continua, where the cubic occurs as the eigenvalue equation of the 3×3 matrices, representing either tensions or deformations. The best known classical case illustrating this statement, and the case which started the whole theory of continua, is the Fresnel theory of light [10], first developed based on the classical Euclidean theory of surfaces. Thereby the light is considered a material structure in a continuous medium (the ether), and Fresnel has shown that its surface - the wave surface - can be synthetized from bits and pieces offered by the diffraction experiments. The physical explanation of Fresnel, which was the basis of the future theory of elasticity, is the existence of some elasticities of ether, as a continuous medium supporting the light (for a pertinent and penetrating review of the case for theory of elasticity in physical optics, the reader ca consult the work of Barré de Saint-Venant [11]). Later on, these elasticities were incorporated in a matrix, whose eigenvalue equation is a cubic.

Now, the Fresnel theory was obviously not enough for describing the light. From a global point of view, it had to be completed up to an affine-type theory by Kummer [12]. For a clear and comprehensive account of the Kummer's theory with significant improvements and additions, one can consult Meibauer's work [13]. The necessity of such a theory is mathematically quite natural, and is nowadays obvious via the affine theory of surfaces: if the ether is an elastic matter, then it can be characterized by a cubic representing its tensions, which then can be taken as the Fubini-Pick cubic of the local affine geometry of the wave surface. The calculation of principal elasticities requires the cubic equation obtained when the Fubini-Pick cubic vanishes. This is in turn a characteristic of the quadric surfaces, thereby mathematically justifying, on one hand, the necessity of Fresnel's hypothesis of the "ellipsoid of elasticities" of ether and, on the other hand, the necessity of considering an affine differential theory of surfaces in physics. Of course, the theory cannot be as direct as presented in these broad lines but, historically speaking at least, the core of the issue is indeed here.

This historical case can serve as a lesson: in the case of a material structure in general, only the constituent particles of the structure can be assumed to be a material continuum, delimited by a surface. The Fubini-Pick cubic form of this surface is then related to the internal characteristics of this continuum, like stresses and strains, of which one cannot talk for the material structure itself without involving the idea of force and of flux of forces.

5.5 The three-dimensional affine transversal space

One further point remains to be secured for an affine differential theory of surfaces, in order to make it properly serve the physical purposes: the dimension of the surrounding space. Starting from the 9th decade of the last century, the immersion of 2-dimensional manifolds in spaces of different dimensions higher than 2 - not necessarily 2+1 as in the classical case above - started attracting the attention of geometers from different perspectives. For the use of exterior differential forms in order to treat such a problem, mention should be made of the pioneering work of Wilkinson [14]. However, we have found a particularly inspirational work strictly referring to the immersion of an affine surface in a five dimensional affine space [15]. This work stimulated an approach of immersion of surfaces in five dimensional space in case of a null Fubini-Pick tensor [16], as well as a generalization, showing a property of minimality for the case of immersion of n-dimensional real manifolds into $n+n(n+1)/2$-dimensional real spaces [17]. The common approach of the problem of immersion here is by considering a general affine frame in the host space, and adapting it to the (hyper)surface in order to induce an affine structure upon it, in connection with that of the space. Considering such a minimality property, the ambient space of a surface should definitely have the dimension five. In order to offer the gist of the matter we shall proceed momentarily more directly though, extending the analogy with the classical theory of surfaces, and only then go a little further, to the physical argument which involves the idea of classical inertia.

First of all, we will consider the classical definition of the displacement as in equation (5.1) but with three "normal" components of the displacements, corresponding to three arbitrary directions in space

$$d\mathbf{m} = s^k \widehat{\mathbf{e}}_k \equiv s^\alpha \widehat{\mathbf{e}}_\alpha + s^{31}\mathbf{e}_{31} + s^{32}\mathbf{e}_{32} + s^{33}\mathbf{e}_{33}. \tag{5.29}$$

This corresponds to the fact that the normal direction to the surface "splits" into three different "directions". The notation is explained by the fact that on the surface one can always find means to define a metric tensor and thereby define further the orthonormality. However, it is not at all a clear thing how the orthonormality on a "normal" direction works. Here one has to rely upon some abstract properties of the environmental space. There are thus quadratic representations of these directions, whereby one can think of representation (5.29) as of a sort of Veronese representation of the affine surface [16]. Anyway, the reference frame should be defined by classical-like equations, written nevertheless in such a way, as to account for the "split" of the normal direction:

$$\begin{aligned} d\widehat{\mathbf{e}}_1 &= \omega_1^1\widehat{\mathbf{e}}_1 + \omega_1^2\widehat{\mathbf{e}}_2 + \omega_1^{31}\mathbf{e}_{31} + \omega_1^{32}\mathbf{e}_{32} + \omega_1^{33}\mathbf{e}_{33}, \\ d\widehat{\mathbf{e}}_2 &= \omega_2^1\widehat{\mathbf{e}}_1 + \omega_2^2\widehat{\mathbf{e}}_2 + \omega_2^{31}\mathbf{e}_{31} + \omega_2^{32}\mathbf{e}_{32} + \omega_2^{33}\mathbf{e}_{33}. \end{aligned} \tag{5.30}$$

This represents the fact that variation of the frame on the surface has both a general in-surface contribution, and a contribution due to the variation of the normal. As to the normal subframe, we maintain the idea that it has only a surface variation, with no normal contribution, for all components:

$$d\mathbf{e}_{31} = \omega_{31}^1\widehat{\mathbf{e}}_1 + \omega_{31}^2\widehat{\mathbf{e}}_2, \; d\mathbf{e}_{32} = \omega_{32}^1\widehat{\mathbf{e}}_1 + \omega_{32}^2\widehat{\mathbf{e}}_2, \; d\mathbf{e}_{33} = \omega_{33}^1\widehat{\mathbf{e}}_1 + \omega_{33}^2\widehat{\mathbf{e}}_2. \tag{5.31}$$

These are just three different copies of the equation (5.6) above.

5.6 A physical argument involving the idea of spin

Theoretically, one can safely argue indeed that the dimension of the environmental space cannot be three, as classically suggested, but the issue is a little more involved, for it cannot be purely geometrical. Of course, one cannot bring arguments against the intuitive feeling that the residence space of the matter, as it presents itself to our senses, is threedimensional. The simplest argument for this statement is provided by the fact that in order to evaluate quantitatively the shape of a body, no matter of the space scale, we need at least three measurements of its dimension along three different directions. However, from the very foundation of the affine theory of surfaces delimiting the matter - the linear independence of the vectors - the things cannot remain at the intuitive level, because the physics itself points out to a environmental space of geometrical dimension five. Indeed, our line of introducing the physics here is based on the fact that both the metric form and the second fundamental form in a point of an affine surface are binary quadratic forms. The binary variable is provided, of course, by the components of differential displacements along the tangent plane to surface in the chosen point. It is from this perspective that the affine normal direction is, in a sense, an affine vector in a three dimensional real space, even though it has to be considered as a direction in space, as we usually know it. The mathematical basis of this statement can be presented as follows.

Any binary quadratic form can be uniquely expressed as a linear combination of three nonsingular, mutually apolar, binary quadratic forms. The proof here is due to Dan Barbilian [18], but for the general theory of apolarity of the binary quadratic forms, one can consult the treatise of Burnside & Panton [9]. The chapter XVI of the second volume, particularly the example 6, pp. 136-137, is particularly illuminating for what we have to say. Start with designating a generic binary quadratic form by

$$Q(x,y) \equiv ax^2 + 2bxy + cy^2$$

and write it as a linear combination

$$Q(x,y) \equiv \lambda_1 Q_1(x,y) + \lambda_2 Q_2(x,y) + \lambda_3 Q_3(x,y),$$

where Q_k represent three quadratic forms whose coefficients are a, b, c with different indices k. This identity is equivalent to the linear algebraical system:

$$a_1\lambda_1 + a_2\lambda_2 + a_3\lambda_3 = a,\ b_1\lambda_1 + b_2\lambda_2 + b_3\lambda_3 = b,\ c_1\lambda_1 + c_2\lambda_2 + c_3\lambda_3 = c. \tag{5.32}$$

Considered as a linear system for the unknowns $\lambda_{1,2,3}$ this system can be compatible, and if the three basic quadratics are nonsingular and mutually apolar, it is always compatible and has unique solution. Indeed, the square of its third order principal determinant can be expressed as

$$\begin{vmatrix} a_1 & a_2 & a_3 \\ b_1 & b_2 & b_3 \\ c_1 & c_2 & c_3 \end{vmatrix}^2 = -\begin{vmatrix} I_{11} & I_{12} & I_{13} \\ I_{21} & I_{22} & I_{23} \\ I_{31} & I_{32} & I_{33} \end{vmatrix},\quad 2I_{kl} = a_k c_l + c_k a_l - 2b_k b_l. \tag{5.33}$$

Now, if the three basic quadratic forms are apolar $I_{kl} = 0$ for $k \neq l$ (see also [7], for the definition of the notion of apolarity of two quadratics), so that only the diagonal entries survive here, and these are the discriminants of the quadratic forms. Because they are assumed nonsingular, the equation (5.33) shows that the determinant of system (5.32) is always nonzero, therefore the system has unique solution for the unknowns $\lambda_{1,2,3}$. If the system is homogeneous, it has only the trivial null solution. Therefore the set of all quadratic binary forms, defined by some external reasons in a point of an affine surface, is a linear threedimensional set.

Likewise, inasmuch as the second fundamental form is to be taken into consideration, the affine normal is not simply a normal direction to a surface, but a transversal linear threedimensional space. If it is to think of an immersion of the affine surface here, the ambient affine space should be therefore five dimensional. This corresponds to the natural fact that in order to completely characterize the profile of a surface in a point, one needs three different perspectives of the surface in three different, but otherwise arbitrary, directions through that point. The curvature of a surface in a point simply cannot be described by a single quadratic form.

Strange as it may seem, the dimension three for the transversal space to a physical surface has indeed strong physical reasons from the very classical mechanics' point of view. That point of view is given by the Newtonian relation between force and acceleration - the second law - within a proper perspective though. In a classical differential geometry of surfaces, as presented above, this perspective amounts to the following. Assume that $\mathbf{v}$ is the local speed of a motion along the geodesics of a surface in a certain point:

$$\mathbf{v} = \dot{s}^1 \widehat{\mathbf{e}}_1 + \dot{s}^2 \widehat{\mathbf{e}}_2.$$

Here the overdot denotes a time rate as usual. This is an intrinsic vector, just like the elementary displacement on the surface itself, from which it is derived. The rate of displacements is taken with respect to the time of geodesics of surface, which is to assume that they exist and the motion along them is physical. The variation of vector v along geodesics has only a component, normal to surface:

$$d\mathbf{v} = (\dot{s}^\alpha \Omega^3_\alpha) \widehat{\mathbf{e}}_3.$$

Using the definition of the curvature vector, one can see that the quadratic generated by the second fundamental form in association with the time of geodesics on surface, i.e. the rate

$$\frac{d\mathbf{v}}{dt} = \dot{\boldsymbol{s}}^\alpha h_{\alpha\beta} \dot{\boldsymbol{s}}^\beta \tag{5.34}$$

is the magnitude of a normal acceleration to surface. By the same token, the intrinsic vector normal to geodesics

$$\mathbf{p} = \dot{s}^2\widehat{\mathbf{e}}_1 + (-\dot{s}^1)\widehat{\mathbf{e}}_2,$$

when transported by paralelism along geodesics, yields a normal vector

$$d\mathbf{p} = (\dot{s}^2\Omega_1^3 - \dot{s}^1\Omega_2^3)\widehat{\mathbf{e}}_3, \tag{5.35}$$

having the magnitude

$$\frac{dp}{dt} = \dot{s}^\alpha h^*_{\alpha\beta}\dot{s}^\beta, \;\; \mathbf{h}^* \equiv \mathbf{h}\cdot\mathbf{i}, \;\; i \equiv \begin{pmatrix} 0 & -1 \\ 1 & 0 \end{pmatrix}. \tag{5.36}$$

This is an acceleration due to the geodesic torsion: the vector (5.35) represents the torsion of any curve touching the geodesic in the given point. The two quadratic forms from equations (5.34) and (5.36), represent two different acceleration, both vectors oriented along the normal to surface. Therefore in a regular geometry they are just two collinear vectors. Nevertheless, as quadratic forms, they are insufficient for characterizing the whole magnitude of the normal acceleration, inasmuch as, considered as a quadratic polynomial, this one is a point in a threedimensional linear space, as shown above. This is the right place to use the property of apolarity in order to properly make up the magnitude of acceleration.

Indeed, as one can see directly, the two quadratics (5.34) and (5.36) are reciprocally apolar. Then we can naturally construct a third quadratic, apolar with each one of them, the so-called resultant:

$$\dot{s}^\alpha h^{**}_{\alpha\beta}\dot{s}^\beta, \; \mathbf{h}^{**} \equiv \frac{1}{2}(\mathbf{h}^1 - \mathbf{h}^2), \; \mathbf{1} \equiv \begin{pmatrix} 1 & 0 \\ 0 & 1 \end{pmatrix} \tag{5.37}$$

where h is the determinant of $\mathbf{h}$, i.e. the Gaussian curvature of surface. If the support function of the surface (see [3] for a clear explanation of the concept) can be represented by a quadratic form, then the proper representation of the acceleration when referred to the time of the geodesics on the surface, is by a linear combination of the quadratics (5.34), (5.36) and (5.37).

As a first conclusion therefore, assuming the classical Newtonian relation between acceleration and force, the forces with which a particle surface responds in accomodating with each other the exterior and interior changes of the matter, are linear combinations of the three accelerations. One can say that the coefficients entering such an expression of the forces represent three types of inertia, which only accidentally reduce to one, in those cases where the force can be sufficiently characterized as vector, for instance when the constituent particles of the matter structure can be represented as a material points. This was actually the case that allowed introducing forces in physics.

Indeed, this was the manner of conceiving forces responsible for the celestial harmony in the first place: by the ratio of their magnitudes (Newton, Principia, Book I, Proposition VII, Corollary 3). Only, in the original case of Newton, the forces were considered as acting in different directions in space, for instance from a planet toward the focus of its orbit, and from the same planet toward the center of the orbit. However, a more explicit example of forces having as magnitude a linear combination of magnitudes of forces, is provided by the case of revolving orbits, whereby the magnitude of the force responsible for the motion along a revolving Kepler orbit is a linear combination of the magnitude of a force inversely proportional to the square of distance and the magnitude of a force inversely proportional to the cube of distance (Principia, Book I, Proposition XLIV). For a treatment of the problem in modern theoretical terms see [19-21].

There is, however, a very instructive classical case illustrating the issue even from the very standpoint of the very theory of surfaces. Like anticipating the case of Fresnel two centuries later, Newton explained the phenomena of reflection and refraction of light by a force acting at the surface of matter, along the normal to that surface (see [22], pp. 79ff). Nowadays, we should be able to say that such a force can be explained within differential geometry, by the extension of the principle of inertia, as long as the light itself is considered as a material structure, as it always was actually. From this perspective, one can say that the diffraction phenomena studied by Fresnel, only added a local structure to the Newtonian force, which was actually based upon the experience regarding only reflection and refraction of light. The existence of such force was indeed confirmed as a physical fact, on one hand through the electromagnetic structure of the light and, on the other hand, through the experiments of P. N. Lebedev [23,24].

It would seem therefore that, by an extended principle of inertia, the magnitude of the force itself is a linear space with the dimension decided by the dimension of the curvature algebra, which is obviously a sl(2,R) algebra. Indeed, the three quadratics above, representing accelerations, can be considered as connected with the homographic action of the three matrices which belong to a basis of $sl(2,R)$ algebra:

$$\mathbf{e}_1 = \begin{pmatrix} -\beta & -\gamma \\ \alpha & \beta \end{pmatrix}, \ \mathbf{e}_2 = \begin{pmatrix} (-\alpha-\beta)/2 & \beta \\ \beta & (\gamma-\alpha)/2 \end{pmatrix},$$

$$\mathbf{e}_3 = \begin{pmatrix} -\beta(\alpha+\gamma) - \beta^2 - \gamma^2 + \Delta & \\ \alpha^2+\beta^2-\Delta & \beta(\alpha+\gamma) \end{pmatrix}$$

The quadratics in question are the symplectic forms corresponding to the linear action of these matrices. As long as we have to deal with a single surface in describing the limit of separation of a particle - in which case α, β and γ are constants - one can argue that this basis is just enough in order to write the physics in a geometric language. However, a physical limit of an extended particle cannot be simply a geometrically fixed surface, but should also contain its local deformations due to physical causes. This situation can only be modeled as a succession of geometrical surfaces, resembling to the known geometrical image of a 1-form for instance (see [25]). This means that the curvature parameters vary in a certain way along a direction crossing the evolving surface, in any of its momentary instances.

In a previous work [2], we have defined such a variable geometry in terms of the second fundamental form of an initial surface in a point, and its different deformations. These deformations are represented by three mutually apolar quadratics which can be interpreted as symplectic forms corresponding to the following three involutive anticommuting matrices

$$\mathbf{e}_1 \equiv \frac{1}{\sqrt{\Delta}}\begin{pmatrix} -\beta & -\gamma \\ \alpha & \beta \end{pmatrix}, \ \mathbf{e}_2 \equiv \frac{1}{\sqrt{\Delta'}}\begin{pmatrix} -\omega_2/2 & -\omega_3 \\ \omega_1 & \omega_2/2 \end{pmatrix},$$

$$\mathbf{e}_3 = \frac{1}{\sqrt{\Delta\Delta'}}\begin{pmatrix} -\Omega_2/2 & -\Omega_3 \\ \Omega_1 & \Omega_2/2 \end{pmatrix}, \tag{5.38}$$

where the entries of the vectors e_2 and $\mathbf{e}_3 \equiv \mathbf{e}_1\mathbf{e}_2$ are exterior 1-forms in the curvature parameters,

$$\omega_1 = \frac{\alpha d\beta - \beta d\alpha}{\Delta}, \ \omega_2 = \frac{\alpha d\gamma - \gamma d\alpha}{\Delta},$$

$$\omega_3 = \frac{\beta d\gamma - \gamma d\beta}{\Delta}, \ \Delta \equiv \alpha\gamma - \beta^2 \tag{5.39}$$

and

$$\Delta \equiv (\omega_2/2)^2 - \omega_1\omega_3.$$

The multiplication table of these three involutive matrices is given by the relations

$$\begin{matrix} \mathbf{e}_1^2 = -1, \\ \mathbf{e}_1 \cdot \mathbf{e}_2 = \mathbf{e}_2 \end{matrix}, \quad \begin{matrix} \mathbf{e}_2^2 = 1, \\ \mathbf{e}_3 \cdot \mathbf{e}_1 = \mathbf{e}_2 \end{matrix}, \quad \begin{matrix} \mathbf{e}_3^2 = 1, \\ \mathbf{e}_2 \cdot \mathbf{e}_3 = -\mathbf{e}_1 \end{matrix} \quad \therefore \quad \begin{matrix} [\mathbf{e}_1, \mathbf{e}_2] = 2\mathbf{e}_3, \\ [\mathbf{e}_3, \mathbf{e}_1] = 2\mathbf{e}_2, \\ [\mathbf{e}_2, \mathbf{e}_3] = -2\mathbf{e}_1, \end{matrix}$$

As vectors, they can represent indeed an orthonormal frame, but with respect to the product

$$2(\mathbf{e}_k, \mathbf{e}_j) = tr(\mathbf{e}_k \cdot \mathbf{e}_j),$$

where the dot means matrix multiplication. For further details regarding such a matrix frame and the inspiring literature, the reader should consult the work cited right above, for our present treatment will turn in some other direction, in order to reveal an inedit classical connection, between inertia and spin.

It is quite clear that the three quadratic forms corresponding to the vectors from equation (5.38) provide us with a basis of three binary quadratics, a, b, c say, in the components of displacement of an initial surface with curvature parameters given by α, β and γ. These quadratics are nonsingular, reciprocally apolar and, if suitably normalized (see especially [18] for a thorough treatment of this issue), they satisfy the quadratic equation [5]

$$a^2 + b^2 + c^2 = 0. \tag{5.40}$$

In other words, these binary quadratics represent in physics what we would like to call the spin parameters. This connection was first outlined by the middle of the last century [26] under "complex space of null radius". In view of our classical inspiration, these parameters should be directly related to inertia properties of the matter. An image then emerges as the basis of description of the free extended particle concept.

An extended particle is thereby aptly represented by its limit of separation from space, which can be organized as a family of surfaces deriving from one another by a process of deformation. One can say that they form a bundle coordinated by the transversal threedimensional space of the curvature parameters. This structure is analogous with a bundle of planes from the classical case of the Yang-Mills gauge fields [27]. In that case a specific choice of frames in the planes would amount to a gauge choice, and the rotation of the frames would amount to a gauge transformation. Here, on the other hand, the construction of the gauge can be explicitly made, and it involves not one, but three parameters. The procedure can be described as follows.

The reciprocal position of two vectors over a surface from the bundle, say e_1 and e_2 is in general not a standard Euclidean one, due to deformation. Once we have a metric at our disposal, it is described by equations of the form

$$\mathbf{e}_1 \cdot \mathbf{e}_2 = \lambda\mu \cos\theta, \ \ \mathbf{e}_1^2 = \lambda^2, \ \mathbf{e}_2^2 = \mu^2, \tag{5.41}$$

where λ and μ are measured quantities. On the other hand, the surface can be characterized by the local orthonormal frame, taken as a standard:

$$\widehat{\mathbf{u}}_1 \cdot \widehat{\boldsymbol{u}}_2 = 0, \ \ \mathbf{u}_1^2 = 1, \ \ \mathbf{u}_2^2 = 1. \tag{5.42}$$

Here $\widehat{\mathbf{u}}_{1,2}$ are the unit vectors of an orthonormal frame adapted to surface. A transformation between the two bases of vectors - properly: a gauging - is given by a nonsingular matrix that reduces the general situation to the standard one:

$$|\mathbf{e}\rangle = \mathbf{m} \cdot |\widehat{\mathbf{u}}\rangle. \tag{5.43}$$

Now, if a, b, c, d are the entries of the matrix m, the equations (5.41), (5.42) and (5.43) imply

$$ac + bd = \lambda\mu \cos\theta, \ \ a^2 + b^2 = \lambda^2, \ c^2 + d^2 = \mu^2. \tag{5.44}$$

These relations determine the matrix a up to an arbitrary phase. Indeed, we can choose

$$a = \lambda \cos\phi, \ \ b = \lambda \sin\phi, \ c = \mu \cos\varphi, \ d = \mu \sin\varphi, \tag{5.45}$$

so that the area determined by the two vectors $\boldsymbol{e}_1$ and $\boldsymbol{e}_2$ is

$$|\mathbf{e}_1 \times \mathbf{e}_2| = \lambda\mu \sin\theta \leftrightarrow ad - bc = \lambda\mu \sin(\varphi - \phi). \tag{5.46}$$

So, if we choose $\varphi = \theta + \phi$, we get

$$|\mathbf{m} = \begin{pmatrix} \lambda \cos\phi & \lambda \sin\phi \\ \mu \cos(\phi + \theta) & \mu \sin(\phi + \theta) \end{pmatrix}. \tag{5.47}$$

This is an Iwasawa representation of the matrix m. Indeed, the matrix is of the form

$$\mathbf{m} \equiv \mathbf{A} \cdot \mathbf{N} \cdot \mathbf{K} \tag{5.48}$$

with the three factors given by

$$\mathbf{A} \equiv \begin{pmatrix} \lambda/(\mu \sin\theta) & 0 \\ 0 & (\mu \sin\theta)/\lambda \end{pmatrix}, \mathbf{N} \equiv \begin{pmatrix} 1 & 0 \\ (\lambda \cos\theta)/(\mu \sin^2\theta) & 1 \end{pmatrix},$$
$$\mathbf{K} \equiv \begin{pmatrix} \cos\phi & \sin\phi \\ -\sin\phi & \cos\phi \end{pmatrix}.$$

Therefore a frame deformation, or even a statistics, which actually comes to the same from a mathematical point of view, is represented by a Iwasawa matrix, which can sustain the theory of a stochastic process for the "skin" of the extended particle, to be described in the manner offered, for instance, by Albeverio & Gordina [28].

The most important point to be noticed here is the mathematical nature of this stochastic process: it is a process with variance given by a metric [29]. This metric is what we would like to call a Barbilian metric for the Riemannian space of families of binary cubics [30]. This can be easiest shown as follows: use the notations

$$\xi \equiv \lambda \cos\phi, \; \eta \equiv \lambda \sin\phi, u \equiv \frac{\mu}{\lambda} \cos\theta, \; v \equiv \frac{\mu}{\lambda} \sin\theta, \tag{5.49}$$

in which case the matrix m can be written in the form

$$\mathbf{m} \equiv \begin{pmatrix} \xi & \eta \\ \xi u - \eta \nu & \xi v + \eta \nu \end{pmatrix}. \tag{5.50}$$

Now, the Killing-Cartan metric of this variable matrix is given by the quadratic form:

$$\left(\frac{dv}{v}\right)^2 - (2d\phi)^2 - 2(2d\phi)\left(\frac{du}{v}\right) \equiv \frac{(du)^2 + (dv)^2}{v^2} - (2d\phi - \frac{du}{v})^2, \tag{5.51}$$

which is indeed the Barbilian metric up to a sign.

It is therefore also important to construct a coframe on a surface of the family. In the parametrization from equation (5.50) the components of this coframe are:

$$\begin{aligned} \omega_1 &= \cos^2\phi \frac{du}{v} - \sin\phi \cos \frac{dv}{v} + d\phi, \\ \omega_2 &= \sin 2\phi \frac{du}{v} + \cos^2\phi \frac{dv}{v}, \\ \omega_3 &= \sin^2\phi \frac{du}{v} + \sin\phi \cos\phi \frac{dv}{v} + d\phi. \end{aligned} \tag{5.52}$$

Thus for a Sasaki parametrization [31] of the local geometry of a surface from the family representing the limit of separation of an extended particle, we have

$$\omega_1 - \omega_3 = \cos^2\phi \frac{du}{v} - \sin 2\phi \frac{dv}{v},$$
$$\omega_1 + \omega_3 = \frac{du}{v} + 2d\phi. \tag{5.53}$$

One can see that $(\omega_2, \omega_1 - \omega_3)$ represents a Bäcklund transformation of the fundamental forms of the Lobachevsky plane, while the form $\omega_1 + \omega_3$ is the corresponding connection form. This explains the form of the gauge metric from the right hand side of the equation (5.51).

Bringing here a standard orthonormal frame on the surface into argument, seems to add a subjective note to the procedure, which in turn brings the question: is this gauging natural? As long as the internal structure of a particle can be taken as a continuum, the answer is definitely affirmative. Indeed, the evolution of the structure of particle is then sufficiently represented by a family of cubics which should vanish on the particle surface, in order to offer the corresponding "eigenvalues" of internal stresses for instance. The geometry just presented is actually a Riemannian geometry of such a family of cubics, as it was first presented [30]. The problem is only the reflection of this geometry in the external forces with which the particle is manifested in a material structure. As a matter of fact, the light is again "illuminating" here: the above gauging is historically related to the electromagnetic structure of light. This issue is presented in detail in one of our previous works (see [32], pp. 148-173)

5.7 Extended particle forces and spin parameters

Assuming that we have an extended particle model, built along the lines sketched above, the problem is to incorporate such a particle within a physical structure of the matter. A first issue in solving the problem, at least in a classical view, is the construction of a potential, and we are able to show here that this potential is strictly determined by the limit of separation of the extended particle. The expression of the force obtained from this potential depends solely on the curvature and its variation. This can be shown as follows.

By the beginning of the last century, Edmund Whittaker was able to produce solutions of the nonhomogeneous Laplace equation (in fact of Helmholtz equation) by introducing arbitrary functions in an integral over the unit sphere [33]. Specifically, if we parameterize the sphere by the polar angles θ, ϕ with respect to its center, then we find that the potential written in the form

$$U(\mathbf{r}) = \oiint_{Sphere} e^{x \sin\theta \cos\phi + y \sin\theta \sin\phi + z \cos\theta} f(\theta, \phi) d\theta d\phi$$

with $f(\theta, \phi)$ an arbitrary function, is a solution of the equation $\Delta U(\mathfrak{r}) = U(r)$.

Afterwards, Pierre Humbert noticed the fact that the partial differential equation to be satisfied by the potential of a surface in space is actually dictated by the equation of the surface itself, if it is represented by a known algebraical expression, as in the case of Whittaker's unit sphere, and that such a determination asks for means depending exclusively on the geometry of surface. This can be shown quite directly, considering $U(r, a)$ the potential between points r and a in space [34]. Humbert chose for illustration a Gaussian central potential, but he soon noticed that the theory goes with arbitrary forms of potential. If the point a describes a surface, S say, the potential of that surface in the space point r is naturally given by the integral

$$U(\mathbf{r}) = \iint_S U(\mathbf{r}, \mathbf{a}) d^2\mathbf{a},$$

where d^2a is the elementary measure on that surface. For a potential of the form

$$U(\mathbf{r}, \mathbf{a}) = e^{\mathbf{a}\cdot\mathbf{r}} f(a), \tag{5.54}$$

if the surface is algebraic, we have

$$\sum_{m,n,p} a^m b^n c^p = 1 \therefore \sum_{m,n,p} \partial_x^m \partial_y^n \partial_z^p U(\mathbf{r}) = U(\mathbf{r}). \tag{5.55}$$

These results generalize the one due to Whittaker, for the representation of the solutions of Laplace equation: in that case it is sufficient to consider that the algebraic equation in (5.55) is a sphere.

The Humbert's results can yet be seen from another angle, namely as an "eigenvalue problem", involving "level surfaces", so to speak. Indeed, if the equation of the algebraical surface is of the general form

$$\sum_{m,n,p} A_{mnp} a^m b^n c^p = \lambda, \tag{5.56}$$

where A_{mnp} and λ are some parameters, then the partial differential equation satisfied by $U(r)$, as given by equation (5.54), is

$$\sum_{m,n,p} A_{mnp} \partial_x^m \partial_y^n \partial_z^p U(\mathbf{r}) = \lambda U(\mathbf{r}). \tag{5.57}$$

This allows us to include among the representations of Humbert some important homogeneous equations like, for instance, the Laplace equation proper. The solutions of this equation are then represented by an integral of the form

$$U(\mathbf{r}) = \iint_S e^{r \cdot a} f(\mathbf{a}) d^2\mathbf{a}, \tag{5.58}$$

where $f(a)$ is an arbitrary function on the surface $a^2 = 0$ (see also [35], Chapter VIII; [36], pp. 18 - 20).

Thus, in view of what has been said before, the potential of a particle in a point from space, considered as a solution of the Laplace equation in that point of space, is decided by the spin parameters. If these spin parameters are characteristic to the curvature parameters and their variations, the force generated by an extended particle in any point in space is offered by the gradient of the potential, as usual:

$$\nabla U(\mathbf{r}) = \iint_S a e^{r \cdot a} f(\mathbf{a}) d^2\mathbf{a} \tag{5.59}$$

a conclusion in concordance with the extended inertia principle as restated above. The arbitrary function $f(\mathbf{a})$ can be given from further invariance considerations, but related to surface only. In a previous work, already cited above [7], we treated such a case by statistical theoretical methods, but the results are in concordance with the group theoretical methods, allowing us to conceive the classical theory of forces as a gauge theory, for the elaboration of which we reserve a future work.

5.8 Perspectives

Any matter structure is theoretically thought in terms of particles. The experience shows that such particles are by no means "elementary", but they exhibit themselves a structure in concordance with the matter structure to which they momentarily belong. In a word, neither that matter structure, nor the structure of the constituent particles are fixed, but are both essentially ephemeral and therefore continuously changing. The classical case of molecules, atoms and nuclei is notorious: none of these structural particles proved eventually to be what they were intended to be. Closer to our times, the modern realm of elementary particles exhibits a wild variety, thus making the term "elementary" obsolete, if it is to be taken within the initial understanding.

Neither the classic nor the quantum theory of forces can properly cope with this emerging image of the world, which seems to us quite natural, and we think that this is mainly due to the fact that the prevailing theoretical image of the particle is that of a material point. This image can indeed be properly used only in the limit of large distances between particles. However, in close particle encounters it obviously fails: there the classical material point needs to be replaced with an extended particle. The present work is meant to build the ground for such a task.

We start from the observation that an extended particle means a finite volume of matter in space, and thus a fortiori a limit of separation of the matter from the empty space. It seems just natural that, when incorporated in a material structure, an extended particle should undergo transformations in order to cope with that structure. This means that its internal evolution should be in concordance with the evolution of the material structure to which it belongs momentarily, and this fact brings the "skin", i.e. the limit of separation of the particle, to the fore. Our declared task here should thus be accomplished by describing this limit of separation from geometrical point of view, with emphasis on the possibilities of introducing the physics into the geometrical theory. We show that the Cartanian approach of the differential geometry of surfaces can properly accomodate the physical needs.

Thus, our main result is that the forces issuing from a particle depend exclusively on the geometry of its limit of separation from the empty space. It is like the forces are actually determined "holographically" - to use a theoretical term nowadays in fashion. As a matter of fact, the holography acquires in our theory a precise meaning, upon which we resolve to elaborate in a future work. For the moment being our results indicate that the classical principle of inertia needs to be extended in order to include three "mass" coefficients, as, for instance, in the old theory of electrons. That theory has partially failed only due to assigning an exclusive electromagnetic nature to the mass, which should not be the case: as historical examples show, the electromagnetic theory is only particular in the realm of forces.

If it is to assign masses to forces by an extended principle of inertia, then these masses cannot be tensors, as in the classical case of the electromagnetic mass, but coefficients connecting the forces to the spin properties of matter which, again, cannot be revealed but within an extended particle model. The masses are simply three coefficients which, if geometrically conceived, cannot be but coordinates in a threedimensional Lorentz geometry. The whole theory of forces is then a gauge theory based on spin properties of matter, of which we presented in this work a representative, based on the following general philosophy.

The structured matter is a collection of extended particles, with forces among them, just as classically conceived. Only, in the classical case, one has material points instead of extended particles. This compels us to consider, first, the internal structure of a particle, which can be dealt with in an affine differential theory of surfaces, as presented in this work. Specifically, the structure of a particle can be sufficiently represented as a homogeneous continuum, the only nonhomogeneity being manifested in the limit of separation of the matter from space. As to the forces between particles, nothing changes with respect to the classical case: they are still represented by the gradient of a potential. Only, when it comes to their calculation, it turns out that they can be calculated from the spin structure generated by the limit of separation of particle from space.

One of the advantages of the gauging theory as presented in this work is its versatility: it can be used in a stochastic approach of the interaction, which is particularly attractive, for instance, in a plasma theory, or in the theory of nanostructures, to name two fields of technological interest today. Furthermore, the explicit connection of the force with the limit of separation of extended particle, limit that can, on occasion, be described via fractals, allows us to entertain the idea that the theory of fractals should be somehow essential here. Indeed, the approach of fractal by the so-called scale theory, is based upon partial differential equations of the kind encountered here in the theory of potential. It then can be shown that a wave theory is instrumental in applying the fractals to physical, or even technological problems, involving matter structures [37]. For other details see [38].

References

1. Poisson, S.D.: Nouvelle Theorie de l'Action Capillaire, Bachelier, Pere et Fils, Paris, (1831).
2. Mazilu, N., Agop, M.: Modern Physics Letters A 30 (2015).
3. Spivak, M.: A Comprehensive Introduction to Differential Geometry, Publish or Perish, Inc., Houston, Texas, (1999).
4. Lowe, P.G.: Mathematical Proceedings of the Cambridge Philosophical Society, 87, 481, (1980).
5. Chern, S.S., Terng, C.L.: Rocky Mountain Journal of Mathematics, 10, 105, (1980).
6. Yau, S.-T.: Proceedings of the American Mathematical Society, 106, 465, (1989).
7. Mazilu, N., Ioannou, P.D., Agop, M.: Modern Physics Letters A, 29, (2014).
8. Cheng, S.-Y., Yau, S.-T.: Communications on Pure and Applied Mathematics, 39, 839, (1986).
9. Burnside, W.S., Panton, A.W.: The Theory of Equations, Dover Publications, New York, (1960).
10. Fresnel, A.: Mémoirs de l'Académie des Sciences de l'Institute de France, Tome 7, 45, (1872).
11. de Saint-Venant, B.: Sur les Diverses Manires de Présenter la Théorie des Ondes Lumineuses. Gauthier-Villars, Paris, (1872). (Extrait des Annales de Chimie et de Physique, Tome XXV (1872), 4e série, 335-381.)
12. Kummer, E.E.: J. Reine Angew. Math. B, 57 (1860), 189-230 (English translation by D. H. Delphenich, www.neo-classical-physics.info).
13. Meibauer, R.O.: Theorie der Gradlinigen Strahlensysteme des Lichts, Berlin, edited by C. Lüderitz'sche. A. Charisius (1864); (English translation by D. H. Delphenich, www.neo-classical-physics.info) (1864).

14. Wilkinson, S.: Mathematatische Zeitschrift, 197, 583, (1988).

15. Decruyenaere, F., Dillen, F., Vrancken, L., Verstraelen, L.: International Journal of Mathematics, 5, 657, (1994).

16. Magid, M., Vrancken, L.: Differential Geometry and its Applications, 14, 125, (2001).

17. Sasaki, T.: Geometriae Dedicata, 57, 317, (1995).

18. Barbilian, D.: Elementary Algebra, in The Didactic Works of Dan Barbilian, Vol. II, Editura Tehnică, Bucharest, (1971) (in Romanian).

19. Chandrasekhar, S.: Newton's Principia for the Common Reader, Oxford University Press, New York, (1995).

20. Lynden-Bell, D.: The Observatory, 126, 176, (2006).

21. Whittaker, E.T.: Analytical Dynamics, Cambridge University Press, Cambridge, (1917).

22. Sir Isaac Newton: Opticks, or a Treatise of the Reflections, Refractions, Inflections & Colours of Light, Dover Publications, Inc., New York, (1952).

23. Lebedef, P.: Les Forces de Maxwell-Bartoli dues a la Pression de la Lumière, Rapports Présentés au Congrès International de Physique, edited by Guillaume & L. Poincaré, Eds, Vol. II, 133-140, Gauthier-Villars, Paris, (1900).

24. Lebedew, P.: Astrophysical Journal, 15, 60, (1902).

25. Misner, C.W., Thorne, K.S., Wheeler, J.A.: Gravitation, W. H. Freeman and Company, San Francisco, (1973).

26. Yamamoto, T.: Progress of Theoretical Physics, 8, 258, (1952).

27. Felsager, B., Leinaas, J.M.: Nuclear Physics B, 166, 162, (1980).

28. Albeverio, S. M. Gordina, M.: Bull. Sci. Math., 131, 738, (2007).

29. Felsager, B., Leinaas, J.M.: Nuclear Physics B, 166, 162, (1980).

30. Levy, P.: Processus Stochastiques et Mouvement Brownien, Gauthier-Villars, Paris, (1965).

31. Barbilian, D.: Comptes Rendus de l'Académie Roumaine des Sciences, 2, 345, (1938).

32. Sasaki, R.: Nuclear Physics B, 154, 343, (1979).

33. Mazilu, N., Agop, M.: Skyrmions-a Great Finishing Touch to Classical Newtonian Philosophy, Nova Publishing, New York, (2012).

34. Whittaker, E.T.: Mathematische Annalen, 57, 333, (1903).

35. Humbert, P.: Sur le Potentiel Correspondant une Attraction Proportionnelle à $\rho \cdot \exp(\rho 2/2)$, Mathematica, Cluj-Napoca, 1, 117, (1929).

36. Bateman, H.: Differential Equations, Longmans, Green & Co., London, (1918).

37. Helgason, S.: Groups and Geometric Analysis, Academic Press, Inc., New York, (1984).

38. Nedeff, V., Lazar, G., Agop, M., Eva, L., Ochiuz, L., Dimitriu, D., Vrajitoriu, L., Popa, C.: Powder Technology, 284, 170, (2015).

39. Mazilu, N., Ghizdovăţ, V., Agop, M.: Role of surface gaugin in extended particle interactions: the case for spin, The Europiean Physical Journal Plus, (2016).

CHAPTER 6

The classical theory of light colors: a paradigm for description of particle interactions

6.1 Asymptotic freedom and holographic principle

There are two contemporary concepts which, coming from the physics of past, carry, in our opinion, a special meaning for the physics of future: the concept of asymptotic freedom and the holographic principle. The first one helped rounding the quantum chromodynamics (QCD) as a science of strong interactions of what are currently considered fundamental material particles. It advocates, roughly speaking, the light-like behavior of the particles involved in strong interactions. The second one, closely correlated to the idea of asymptotic freedom, but aiming to cover the whole range of fundamental forces, advocates the idea of two degrees of freedom in the description of fundamental interactions. This, again, is a fundamental classical characteristic of light.

These two modern principles stand witness to the fact that the analogy in theoretical physics is still the main tool that works at any level, in any physical theory. Taken as such they would then point out that the light should be part of a universal model in the theory of fundamental interactions of matter, and thus it should become a standard in the modern analogy leading to fundamental forces. We aim to show here that both these principles hold indeed a manifest classical character, whose roots are to be found, with equal chances, both in the classical Newtonian theory of light and in the quantum theory of light as it was constructed by Planck at the beginning of the last century. In other words, the two modern principles carry the burden of continuity with respect to the two physical theories considered nowadays the quintessence of the classical theoretical physics.

Along the analysis to be presented here, it will become clear that the light is indeed a model of any interaction in the universe, inasmuch as it can be revealed by an obvious interaction: the one giving the colors. This interaction has an exquisite classical description, involving only two degrees of freedom, and that description can be done by continuous Lie groups from the SL(2,R) family. There are thus Yang-Mills type fields describing the colors, in the spirit of the classical theory of light established by Hooke and Newton. This turns out to be also a characteristic of light from the quantum mechanical point of view.

6.2 The Classical Theory of Light: an Abridged History

First, let us review the classical theory of light, up to the point where it was brought by Hooke and Newton. Following the line of thought given by the idea of analogy, we primarily find the standard of the classical analogy which led to the theory of light itself: this is given by the waves on the surface of quiet water, induced by a stone falling into water. One can say that the making of the theoretical model of light is nothing more than elaborations, at any of its significant moments, on one or another of the parts of that experience. The phenomenology which was the basis of these elaborations for the first theoretical model of light comprises two basic observable facts: reflection and refraction of light. Let's review the original significant moments of the making of concept of light.

First, it was the explanation of the propagation of light. To the extent where it can be explained as a motion, that classical standard of analogy shows that it is a "motion of a motion", so to speak, but not only that. The concentric water waves show indeed a motion perpendicular to the water surface, which nevertheless "multiplies", in order to allow for the growth of the circles representing waves. The propagation is therefore a "motion of the periodic motion" but, being accompanied by a continual growth of the waves, it indicates also a continual generation of that periodic motion, along the very circles representing the crests of waves.

In the case of light, though, the analogous of water surface is missing, to say nothing to the effect that, as a matter of fact, even the analogous of the water is missing. To make up for the want of an equivalent to water, the ether was invented: the light waves are waves in ether. However, the water surface was hard to replace, for the propagation of light takes place in all directions but in space not in a plane, and apparently there is no physically distinguished surface to play the part of an equivalent of water surface, like a space "section" as it were. The Huygens theory of the wave surface of light simply supresses the need of that equivalent of water surface, going directly over to space matters, with the help of the concept of light ray.

In space the analogous of circles depicting the waves on the surface of quiet water, was first the so-called "orb", defined by Thomas Hobbes as the material portion between two concentric spheres of close radii, which extends continuously by propagation [1]. What really matters in this definition is only the uniform continuous growth of the sphere, equivalent to the growth of circles on the surface of water, thus giving what seems to be the essence of the propagation phenomenon of light in space. One therefore loses, by simple suppression from the model thus foreshadowing the later principle of Huygens the local oscilatory motion characterizing the waves on the face of quiet water. As such, nothing assures us here that the motion of light within orb would be a vibratory motion. On the contrary, what seems to be obvious on logical grounds is that such a motion is rather uniform, matching the uniform extension of the orb in space. Further on, Hobbes also defines an extremely important concept that played a fundamental part in the development of the whole theory of light. Disregarding even the evidence of the geometrical divergence of the light rays, he defined the so-called light line, which would represent what happens within the surface of water therefore along the very circles depicting the water waves. This led implicitly to the concept of physical ray of light, which is a plane figure formed by two parallel straight lines ideal geometric rays delimiting the physical ray joined perpendicularly, in each one of their corresponding points, by the light line. The propagation of light could now be defined as a transportation of the light line, and thus it gained some other, more subtle nuances, than the standard given by water waves.

First, the lack of a detailed explanation of the differential properties of this light line started to be noticed. We have reasons to say that to Hobbes the propagation of light is simply the parallel transport of this segment in the Euclidean sense, therefore a transportation which maintains the property of orthogonality with respect to limiting rays all along the physical ray. As long as the light goes straight there are no problems. We have to face them only in the case of change of direction of light, like in reflection and refraction phenomena. Here the transport of the light line must be defined accordingly, and indeed Hobbes does it by methods prefiguring those which established the parallel transfort of the modern differential geometry [2]. As Hobbes conceives the transport in propagation of light only within classical synthetic geometry, the problem of extension of the light segment within the orb which is a problem of deformation proper in modern views remained in suspension. It obviously popped up later on, requiring specific solutions related to the idea of gauge.

At this moment the important intervention of Robert Hooke led to the first rational idea of color, which appears in the case of interaction of light with matter, as represented by refraction and reflection phenomena. Apparently Hooke defines the color by noticing first of all that the physical ray of Hobbes is actually an ideal case. Indeed, in reality the geometrical rays delimiting a physical ray of light should certainly be divergent. They can only be approximately parallel, the approximation being closer to the ideal situation as the distance from the light source grows [3, 4]. Therefore the light line of Hobbes is in reality not perpendicular to the rays delimiting a physical ray of light. However, at great distances from the source of light they can be considered very nearly perpendicular, which in practice is quite sufficient. To Hooke the light line acquires even a special name orbicular pulse which suggests the idea of motion characterizing the light within the orb. The idea of pulse would have been suggested to Hooke by a study of Hobbes, who describes the creation of light as a pulsating phenomenon of the kind of systole and diastole, observed in the case of heart ([5, 6]). However, this "orbicular pulse" means a great deal more in the hands of Hooke, along the idea of representation of the way in which the light motion takes place within the orb.

For, as we have noticed, the idea of propagation in space has suppressed the oscillatory motion, which, regardless of its direction, is essential in the clasical standard of analogy that led to the concept of light: the circular waves on the face of quiet water. Even the idea that some motion, other than that directly representing the propagation per se, might be associated with the light was momentarily lost: the light line of Hobbes is a purely geometric concept after all! Hooke calls back into question the problem, and solves it once and for all: within the orb we have definitely motion; the problem is to decide what kind of motion this is. And he reaches, speculatively, the epoch-making conclusion, that the motion characterizing the light within orb is vibratory, therefore an oscillatory motion, like that of the waves on water, perpendicular to its surface. Hooke's reasoning is quite simple, speculating on the fundamental idea that the light is material. In broad lines, one could say that he just notices that if the light would be characterized by a continuous motion within a body like the Hobbes' orb this would lead to rupture, as it surely happens in the processes of continual deformation of materials. As, however, no transparent body is apparently destroyed by the light passing through it, one should conclude that the motion characterizing the light is vibratory, with a very small amplitude (short to Hooke), reestablishing periodically the matter in every one of its points. This motion is localized along the light line of Hobbes so that this one is actually an "orbicular pulse", approximately perpendicular to the geometrical rays defining the physical ray.

Here Hooke introduces the first rational concept of color ever, and thus begins the modern history of theoretical chromodynamics, as understood *mot à mot* according to its Greek name: the dynamics of color. First, the color is not revealed but only by light, when it touches the material bodies, i.e. by interaction. According to Hooke, the color is explained by the fact that through refraction the orbicular pulse acquires an inclination on the limiting rays, becoming slightly nonorthogonal to the geometrical rays, to a specific extent though, depending on the color carried by them. Due to this, in a natural ray, the orbicular pulse is broken into segments, each one representing a homogeneous ray of specific color. So, for the natural light, the orbicular pulse appears as a broken, even folded line, joining the defining geometrical rays, which carry two fundamental colors red and blue from which all the other colors are constructed by the action of matter upon light.

This idea the first true physical theory of colors was killed in the cradle, so to speak, by Newton's observations to the effect that the orbicular pulse is actually a cross-sectional surface phenomenon. It is indeed performed transversally with respect to the direction of light ray, but not in a definite plane containing that direction. In other words, the color is indeed a parameter of homogeneity of the physical ray of light, but spatially not planar. A physical ray of light of a given color is symmetrical around the direction of propagation - in modern terms it has axial symmetry [7]. So if one speaks about Hooke's fragmentation of the orbicular pulse, with each fragment representing a color, this process does not preserve the axial plane of the original pulse in the transversal section of the physical ray. The physical ray itself is therefore not a plane figure, as Hobbes and Hooke presented it, but has a volume as the experience on light shows. If the light carries many colors, for instance if it is white, then it is transversally inhomogeneous, i.e. it is no more axially symmetric, although not as geometrical shape, but with respect to color. As such, it is indeed composed from axially symmetric color-homogeneous rays, that can be exhibited as an elongated spectrum, when passing the light ray through a prism. This is, in broad strokes, the newtonian conclusion of the celebrated prism experiments, and the ground for Newton"s discussion on colors, which later on, during 19th century, generated the physical theory of light spectrum, and implicitly led to the theory of quanta.

This moment of our knowledge has a particular significance when judged through the formulation of the modern holographic principle [8, 9]. Indeed by the physical ray of Hoobes and Hooke, the idea of planeness, as contained in the waves on the water, is certainly preserved. Newton"s intervention sets things in order according to the observations on the light itself. Formulated in modern terms, the Newtonian conclusion is: the physics of light itself shows that the planeness is not preserved in the geometrical form of the physical ray of light, but in a general property of symmetry, abstract we might say, that can be expressed in two variables. In other words two degrees of freedom suffice to describe the light! Let's elaborate a little more on this statement.

Quite obviously, Newton realized that, from the point of view of the experience on light itself, something is missing in the concept of physical ray of Hooke. That concept cannot explain the fact that the experimental white light ray is spatially homogeneous and isotropic in the cross section, when the ray is constructed by appropriate circular holes. According to Hook's idea, the cross section should appear on a screen differently colored on a certain direction of that screen. It appears this way indeed, but only when one complicates the construction of the ray by passing the light through a prism. Only this procedure isolates homogeneous, axially symmetric, "sub-rays" of the same color, in terms of which one can explain the elongated spectrum in the manner of Hooke. This is, however, a general idea of symmetry, and it does not refer by any means to the geometrical shape of the ray. In other words: it is the variation of light color that has two-dimensional extension, not the ray itself. Thus, one can say that the study of light per se adds this important conclusion to the very idea of wave, above and beyond the classical standard of analogy that helped constructing the concept thus far.

Another important point in Newton's observations is the one usually connected with the particle theory of light. True, sometimes and quite often at that, one might say Newton slips into the direct association of light with material particles, but his definition of the light ray, which starts the celebrated Opticks, is extremely cautious and does not reduce by any means to the idea of particle in the classical connotation. So much the less reduces it to the geometrical concept of straight line. It certainly pays to reproduce here that definition [7], for it comprises a whole philosophy, which later, with the occasion of quanta of light, was labelled as "revolution":

> By the Rays of Light I understand its least Parts,
> and those as well Successive in the same Lines,
> as Contemporary in several Lines.

A consideration of the subsequent text of Newton, explaining this definition, shows two further essential points of the Newtonian natural philosophy. First, Newton refers the definition of "parts" to the experimental possibility to build them: the "least Parts" should not, by any means, be understood as "particles" in the classical connotation. These last ones can exist "under our eyes" as it were, being stable at the time scale of the common experience. They do not require experimental action in order to be defined. On the other hand the "least Parts of light must be defined accordingly, therefore experimentally. And their definition depends on the experimental capability to exhibit them. On one hand one has to have the physical possibility of discerning a direction of propagation of light, by holes in screens for instance, and on the other hand one should have the physical possibility to stop the light by a screen. It is the interplay of these two experimental procedures, obviously in an ideal theoretical form, that defines the least parts of light, the way Newton conceives them.

Therefore, coming again to the modern theoretical environment, Newton defines the rays of light in the modern manner in which the elementary particles are defined: only by experimental capability. Indeed the modern elementary particles are closer to the least parts of light as defined by Newton, than to the matter as intuitively understood. It is in this sense that one can talk indeed of a particle theory of light to Newton. But then one can demonstrate and we will make this obvious in the present work that as such the light can be theoretically considered the paradigm of any theory of fundamental interactions. The discovery of the asymptotic freedom just shows it. Of course, in order to do this, one has to use an explicit holographic principle.

A second point made explicit by Newton himself is that the idea of geometrical ray as a straight line is directly connected with the infinite speed of light. As, however, the experimental evidence of the epoch showed that the light has a finite speed, he declares that he was forced to concede to his definition of the ray reproduced above. It is here an essential point of the natural philosophy of light, that certainly needs elaboration and, we have to confess, we were unable to find it properly documented to Newton. The statement, of course, can be proved within the modern differential geometry, and constitutes one of the most subtle points of the wave-particle transition, seen, again, as an "asymptotic freedom".

6.3 Classical Light in Terms of Two Degrees of Freedom

In this section we present the Hooke's and Newton's results as geometrical theorems. Obviously, the classical theory of light is naturally correlated with the classical differential geometry of surfaces, by the very concept of wave surface. Thus a light ray in vacuum, for instance, can be imagined as a trajectory, in some relation with the local normal of the wave surface. The propagation of light, taken in the initial connotation as the process that should include naturally the reflection and refraction phenomena, can be imagined, first and foremost, as a variation of the normal direction to the evolving wave surface. Of course, with this concept we took a leap over time, coming closer to modern ones, where the Hobbes"? idea of orb metamorphosed into that of wave surface, by bringing in the difraction among the experimental facts characterizing the light per se [10]. However, the presentation of the development of concept in its historical continuity would not be helpful here, inasmuch as we follow a logical continuity, which can only be exhibited by a theory of colors. In order to reach that logical continuity Fresnel's theory had to be rounded, so to speak, by the idea of gauge, which, as we will show here, can be naturally supported only by such an idea of color, leading directly to a Yang-Mills type theory. And, of course, this idea can in turn be naturally extended to characterize the modern concept of particle. Only the classical theory of light, as briefly reviewed in the previous section, would naturally allow for such an extension, and the present section shows the way to it.

The classical differential geometry of the surfaces is entirely constructed on a general manner of conception of the notion of neighborhood [11] . Any point of a surface belongs to a neighborhood of a certain order of another point. Between the neighborhoods of different orders, in the same surface, there is a connection, uniquely characterizing the surface. In order to see how this philosophy works, assume a formal Taylor expansion of the position vector on the surface, something of the form:

$$\boldsymbol{r} = \boldsymbol{x} + d\boldsymbol{x} + d^2\boldsymbol{x} + d^3\boldsymbol{x} + \dots$$

This formula suggests the orders of neighborhoods through the natural orders of the differentials of the positions and . For instance, the second differential is the variation of the first differential, therefore the differential geometry on a first order neighborhood is given by vectors that are second order differentials. The third differential is the variation of the second differential, therefore it reflects a differential geometry for which the second order neighborhoods are the basic "showground", etc. However, it turns out that there is really only one essential "showground", namely the first order neighborhood, inasmuch as the whole geometry in a point of the wave surface, and the attached physics of course, can be described in terms of the first order differentials.

First, one can recognize the departure of $\boldsymbol{r}$ from the surface at the location $\boldsymbol{x}$ by the projection $\boldsymbol{r} - \boldsymbol{x}$ along the normal to surface at $\boldsymbol{x}$, whose unit vector we denote $\widehat{n}$ as usual. We have

$$\widehat{\boldsymbol{n}} \cdot (\boldsymbol{r} - \boldsymbol{x}) = \widehat{\boldsymbol{n}} \cdot d\boldsymbol{x} + \widehat{\boldsymbol{n}} \cdot d^2\boldsymbol{x} + \widehat{\boldsymbol{n}} \cdot d^3\boldsymbol{x} + \ldots \tag{6.1}$$

This means that the vector $(\boldsymbol{r} - \boldsymbol{x})$ does not belong to the surface in all of its local neighborhoods, but has departures of different orders from it, and these become apparent as we get closer to the surface in order to be able to distinguish its details. As by the very definition we have to admit that the vector $d\boldsymbol{x}$ is within the surface, the first term in equation (6.1) is null, so that

$$\widehat{\boldsymbol{n}} \cdot (\boldsymbol{r} - \boldsymbol{x}) = \widehat{\boldsymbol{n}} \cdot d^2\boldsymbol{x} + \widehat{\boldsymbol{n}} \cdot d^3\boldsymbol{x} + \ldots \tag{6.2}$$

Therefore, the first-order neighborhood of the surface is characterized by this differential relation. The length of the first-order differential of the position vector is the first fundamental form of the surface, or the metric. The right hand side of the equation (6.2) represents details that become obvious as we gradually approach the point of the surface. The first term in equation (6.2) is the second fundamental form of the surface. In the intrinsic geometry of the surface the invariants related to the second fundamental form are different measures of the local curvature of the surface.

We don't think that we can escape, in any physical problem, and so much the less in the case of light, from the bounds of analogy altogether. Here it comes with the idea of the coarsening of the wave surface due to matter. This is a phenomenon of "fragmentation" of surface, making it have a certain degree of "roughness". The "roughness" is reflected in the local variation of normal unit vector of the surface, and this is exactly what happens in the cases of reflection and refraction of light analyzed by Hooke and Newton. And, as well known, the variation of that unit normal is a vector within the surface, whose components are two differential 1-forms, designated here as and , which we call the curvature differential forms [12]. Either this vector, or one related in a certain way to it, should be connected with the "orbicular pulse" of Hooke. Thus Hooke idea of representation of the colors by an angle with respect to the position vector has, in the classical theory of surfaces, quite a natural representation.

Now, in ordinary physical situations, the idea of roughness, is naturally connected with that of a friction force: the roughness is variable with the "friction" mechanically representing the interaction between matter formations. The "friction" is essentially a surface phenomenon. The "friction" force is usually zero whenever the surface is smooth, a condition characterized, in a first instance, by the lack of variation of the unit normal to surface, therefore by the fact that the curvature differential forms are null. According to one of the Cartan lemmas [13], this "friction" force, considered as a surface force, should be a differential 2-form, which can be written as

$$f = \omega_1^3 \wedge \phi^1 + \omega_2^3 \wedge \phi^2, \tag{6.3}$$

where ϕ^1 and ϕ^2 are two conveniently chosen differential 1-forms, and "$\wedge$" means exterior multiplication of the differential forms. The equation (6.3) incorporates the previous logic, according to which the force is zero whenever there is not geometrical roughness, i.e. there is no variation of the normal to surface. However, this is only a necessary condition.

If the conveniently chosen auxiliary forms are the components of the first fundamental form, i.e. the components of the position vector $d\mathbf{x}$ in the tangent plane, s_1 and s_2 say, then equation (6.3) simply offers the definition of the curvature matrix as a limiting case where the "friction" forces are zero. Indeed, in the case of null "friction" force in equation (6.3), another one of the Cartan lemmas shows that we must have

$$\begin{pmatrix} \omega_1^3 \\ \omega_2^3 \end{pmatrix} = -\begin{pmatrix} \alpha & \beta \\ \beta & \gamma \end{pmatrix} \cdot \begin{pmatrix} s^1 \\ s^2 \end{pmatrix} \tag{6.4}$$

with α, β, γ some parameters - *the curvature parameters.* This defines the classic symmetrical curvature matrix. The curvature parameters are thereby external parameters, a feature bestowed upon them by the Cartan lemma. They are always submitted to variation due to some physical causes contained in the matter interacting with light. Their geometrical evaluation can be made just by measurements on the surface as usual [15].

If, however, the "friction" force is permanent, like in the case of light through ether, i.e. it always exists and is nonzero by some physical reasons rather than geometrical, we can express it as a differential 2-form:

$$f = \Phi_{\alpha\beta} s^\alpha \wedge s^\beta, \quad \alpha, \beta = 1, 2.$$

Then the equation (6.3) can be read in the form

$$(\omega_1^3 - \Phi_{12} s^2) \wedge s^1 + (\omega_2^3 - \Phi_{21} s^1) \wedge s^2 = 0.$$

Applying again the Cartan lemmas' considerations, we have, in a compact "Dirac" writing

$$\langle\omega^3| - \langle s|\cdot\mathbf{\Phi} = \langle s|\cdot b \therefore \langle\omega^3| = \langle s|\cdot(\boldsymbol{b}+\mathbf{\Phi}),$$

where $\mathbf{\Phi}$ is the 2×2 skew-symmetric matrix having the unique element $\Phi_{12}\equiv\phi$, $\mathbf{b}$ is the usual curvature matrix from equation (6.4) and $|s\rangle$ is the column vector from that equation. We do recognize here a curvature matrix which is no more symmetrical, but contains also a "twisting" naturally accompanying the "roughness" of surface. From among the usual measures of the curvature, the mean curvature of the surface remains unchanged by such physical forces, however its Gaussian curvature is changed:

$$2H \equiv tr(\mathbf{b}+\mathbf{\Phi}) = \alpha+\gamma,\;\; K\equiv \det(\mathbf{b}+\mathbf{\Phi}) = \alpha\gamma-\beta^2+\phi^2.$$

Such forces do not change the second fundamental form of the surface per se. In the case of light, they cannot appear therefore when the propagation of light is made without refraction or reflection, but should nevertheless be obvious in like processes. This fact validates, to a large extent, the electromagnetic image of light, even within the limits of the classical geometrical optics. It has, however, some other, more fundamental connotations related to the wave surface, which become apparent if we take into account the classical considerations of Hooke and Newton.

Indeed, let's just consider the second fundamental form. It is represented by the first term in equation (6.2) which, in view of the orthogonality between the normal to surface and the elementary displacements within surface, can be written as:

$$\widehat{n}\cdot d^2\mathbf{x} \equiv -d\widehat{n}\cdot d\mathbf{x} = \langle s|\mathbf{b}|s\rangle. \tag{6.5}$$

In this expression we used the previous equation (6.4). This tells us that the second fundamental form of the wave surface, being a measure of the projection of the variation of normal unit vector to the displacements within the wave surface, can be taken as a measure of the color of light in the sense of Hooke. The orbicular pulse should therefore be taken in relation with the wave surface, according to Newton's conception of light. This will justify the modern idea of three-dimensionality of the manifold of colors: if the second fundamental form of the wave surface represents the color of light, then the color space is certainly a linear threedimensional space, as claimed in the modern theory of colors [16]. Indeed, one can prove that the binary quadratics do form a linear threedimensional space.

The variation of second fundamental form can generally occur by both the variation of the curvature parameters and by the variation of the components of the first fundamental form in the tangent plane of the wave surface. In a compact "Dirac notation" this variation can be written as

$$\delta II = \langle ds|\mathbf{b}|s\rangle + \langle s|d\boldsymbol{b}|s\rangle + \langle|\boldsymbol{b}|ds\rangle. \tag{6.6}$$

Here II denotes the second fundamental form. We assumed that, in expressing this variation, the rules of usual differentiation apply. There are, therefore, two main contributions to the variation of the second fundamental form in this approach. One of them is due to the variation of the components of the first fundamental form in the tangent plane, the other is due to the variation of the curvature matrix itself. This last variation is the one we are after, for it represents the variation of the second fundamental form strictly due to some physical reasons, for instance the interaction of light with the material environment. The question arises therefore: when is the variation of the second fundamental form strictly due to the interaction with the medium hosting the light?

At least formally, the answer is quite obvious from equation (6.6): in those cases in which its variation is due strictly to the variation of the curvature matrix, i.e. when in equation (6.6) only the middle term remains. This can happen in cases where the sum of the two extreme terms from the right hand side of (6.6) is zero:

$$\langle ds|\mathbf{b}|s\rangle + \langle s|\mathbf{b}|ds\rangle = 0. \tag{6.7}$$

Assuming now the existence of an "evolution" of the vector $|s\rangle$, such that

$$|ds\rangle = \boldsymbol{a}|s\rangle, \tag{6.8}$$

where $\boldsymbol{a}$ is a real matrix, the condition (6.8) becomes:

$$\langle s|(\boldsymbol{a}^t\boldsymbol{b} + \boldsymbol{ba})|s\rangle = 0.$$

Here the superscript "t" stands for "transposed". In other words, when the position in the first order "showground" of a point of the wave surface evolves according to equation (6.8), then the second fundamental form varies strictly due to the curvature parameters if

$$\mathbf{a}^t\mathbf{b} + \mathbf{ba} = \mathbf{0}. \tag{6.9}$$

A solution of this equation presents itself immediately in the form $\boldsymbol{a} = \boldsymbol{I} \cdot \boldsymbol{b}$, where $\boldsymbol{I}$ is the fundamental skewsymmetric 2×2 matrix. The condition (6.9) is satisfied in view of the symmetry of $\boldsymbol{b}$ and the antisymmetry of $\boldsymbol{I}$. The projection of the vector $|ds\rangle$ from equation (6.8) along vector $|s\rangle$ is, in that particular case, the geodesic torsion of the surface. Incidentally, the condition (6.10) is satisfied also for the inverse of the matrix $\boldsymbol{a}$. Then, if the curvature parameters are constants, the evolution (6.8) preserves the second fundamental form of the surface. In that particular case one can properly call the second fundamental form a wavelength. This is the classical way to description of color, leading, through the dispersion law, to the modern theory of coherence.

6.4 Nonconstant Curvature: the Case of Light Line Deformation

What about the cases when the curvature parameters are not constants? The physics of light is then dictated by both the current values of the curvature parameters and their variations. In the first order neighborhood of a certain point of the wave surface, the variations of the curvature parameters can be described by the deformations of the surface. This concept was left behind by Hooke his orbicular pulse is strictly a periodic motion and has not been even considered by Newton. Let us consider it here within the classical differential geometry of a wave surface.

In the case of a symmetric curvature matrix, the equation (6.8) can be integrated even without considering a time parameter, thus leading to ellipses on the wave surface as dictated by constant distance from the tangent plane. This is in fact the classical case of interpretation of the second fundamental form of a surface. One can say that, in case we are able to discover a time, the motion dictated by equation (6.8) is a harmonic two-dimensional motion: just as Hooke claims for his orbicular pulse. However, the orbicular pulse should not be only this displacement. Indeed, as we already noticed, Hooke did not take into consideration the inherent deformation accompanying the expansion of wave, and one might even add that he could not do it. Yet, within the differential theory of surfaces the deformation of the orbicular pulse comes as quite a natural concept. We will illustrate this statement with the help of the classical idea of infinitesimal deformation of wave surface [14].

A deformation of the wave surface is infinitesimal if the first fundamental form in a point coincides with the first fundamental form in the corresponding point of the deformed surface. This definition mimicks what Hooke assumes happening inside transparent bodies penetrated by light. If the deformation can be expressed by a small parameter, say ε, in the form

$$\boldsymbol{r} = \boldsymbol{x} + \varepsilon \boldsymbol{z},$$

then the first fundamental forms at $\boldsymbol{r}$ and $\boldsymbol{x}$ coincide when ε is infinitely small. This sets important restrictions upon vector $\boldsymbol{z}$, amounting to the fact that its differential should be always perpendicular to $d\boldsymbol{x}$ in the first order "showground". If this condition is expressed by a relation of the form

$$d\boldsymbol{z} = \boldsymbol{y} \times d\boldsymbol{x}, \tag{6.10}$$

then the vector y cannot be quite arbitrary: its component along the normal to surface needs to be constant, while the in-surface components have differentials that should be expressed linearly in the components of the fundamental 1-form:

$$\begin{pmatrix} -v^2 \\ v^1 \end{pmatrix} = \begin{pmatrix} A & B \\ B & C \end{pmatrix} \cdot \begin{pmatrix} s^1 \\ s^2 \end{pmatrix}; \quad \begin{pmatrix} v^1 \\ v^2 \end{pmatrix} \equiv d\boldsymbol{y}. \tag{6.11}$$

Then, as a consequence of the fact that $d\boldsymbol{y}$ is an exact differential vector, the parameters A, B, C must satisfy, besides other restricting differential relations, an algebraic apolarity with respect to the second fundamental form of the surface. It is this condition then, which is expected to prevail whenever the vector $\boldsymbol{y}$ is dictated by external circumstances of a physical nature imposed during the process of deformation of surface, because these circumstances are external and local. One should thus have:

$$v^1 \wedge \omega_1^3 + v^2 \wedge \omega_2^3 = 0 \therefore \alpha C + \gamma A - 2\beta B = 0. \tag{6.12}$$

For an algebraic and thus, in fact, physical interpretation of this result, let"s notice that, because $d\boldsymbol{y}$ is an "intrinsic" vector with respect to surface (its component along the normal to surface is null), the cross product of this vector with the elementary displacement on surface is oriented along the normal to surface. This vector is

$$d\boldsymbol{y} \times d\boldsymbol{r} = \{A(s^1)^2 + 2Bs^1s^2 + C(s^2)^2\}\widehat{\boldsymbol{n}}. \tag{6.13}$$

Consequently its magnitude is a quadratic form, algebraically apolar to the second fundamental form of the surface. One can therefore say that it is added to the second fundamental form, thus changing the local curvature of the surface. According to Hooke"s idea it changes the color of light. As the parameters A, B, C, just like α, β, γ, are introduced by external reasons they can be taken as representing the matter acting upon light. Is this image reasonable?

Within classical phenomenology, matter interacts with light only because of its space extension. Incidentally, one could say that, in the modern times, only the idea of material point made possible the electromagnetic theory of light. But, continuing the reasoning within the classical phenomenology, based on reflection and refraction experiments, an extended part of matter has naturally a surface separating it from the environment. The coefficients A, B, C can then be taken as representing the surface of matter from the very same point of view which refers to the wave surface, as described above, i.e. they are the coefficients of the second fundamental form of the surface of the matter interacting with light. In this situation the equation (6.13) carries an important dynamical connotation, occuring in the form of a theorem which appears as a working hypothesis to Newton.

The corollary of this whole classical theory would be that the matter itself behaves like light, inasmuch as it interacts with light through its surface, as it is actually the case. The quantitative expression of this interaction is then the bilinear form from equation (6.13) which is zero only in the case of infinitesimal deformations. The point is that the matter extension can be geometrically described just like the extension of light, by a second fundamental form, representing a color. This, again, corresponds to the natural fact that the color only appears at the interaction of the matter with the light. But then, a normal acceleration occurs at the interaction point, for the second fundamental form is a measure of an acceleration normal to surface (see equation (6.5) above). Or else, as in the case of matter acceleration means inertia, therefore, according to the classical principles, it means a force. Whence, the idea that the matter acts upon light with a force along the normal direction to its surface. This is a fundamental hypothesis to Newton, used by him to prove the laws of refraction [7] . As it turns out, it is actually a geometrical theorem.

6.5 Quantum Theory of Light

The classical theory of light, as extracted from the phenomenology comprising the experiments of reflection and refraction, is not the only one pointing explicitly to a holographic principle. The same happens with the quantum theory of light [15,16], whereby the two degrees of freedom appear to a more abstract level, while having also a more involved physical meaning. Indeed, from a purely statistical theoretical point of view, the Planck moment in the physics of light [17] reveals two distinct theoretical sides. The first one is the heuristic side, closely related to the Gaussian aspect of the statistics of light fluctuations. According to Max Born, this was the source of inspiration in establishing the famous connection between the fluctuations of the spectral density and the equilibrium temperature of the blackbody radiation, leading to the idea of quantum. The second side of the Planck moment of the physics of light is the proper quantum side, whereby the probability distributions characterizing the blackbody radiation are of the quadratic variance function type. The quantum is required here by the condition that the blackbody radiation spectrum should satisfy the Wien displacement law. The contemporary theory of light colors, and of colors in general, seems mainly related to the first side of this moment of physics. Let"s show a way to it

Now, the two degrees of freedom involved in the holographic principle are more intricate, yet closer to the proper holographic description of light. Indeed, the Planck"s original Gaussian, represents two processes of fluctuation, at low and high temperatures, and is uncorrelated [18]. When considered, however, in the general, correlated form, the probability density of this Gaussian would be something of the form:

$$p_{XY}(x,y) = \frac{\sqrt{ac-b^2}}{2\pi} exp\{-\frac{1}{2}(ax^2+2bxy+cy^2)\}, \tag{6.14}$$

where X and Y are, as we said, the two characteristic fluctuation processes, playing the part of the two degrees of freedom, originally constituting the thermal light at low and high temperature, respectively. The classical theory of color can be constructed on this statistics, as it has an interesting twist on it.

Indeed, in the classical theory of color, we don"t specify these two random processes by temperature regimes, because in general we cannot associate a physical temperature with the color. The problem of associating a temperature to the color was not solved yet [19], and we don"t think will ever be solved. For once, the thermodynamically defined absolute temperature is not physically supported for light as classically defined. This issue led to Planck theory in the first place. On the other hand, from a statistical point of view, the temperature goes into a parameter characterizing the distribution of colors in a more elaborate way than it does in the Planck's statistics. Thus, let"s just say, for the sake of the present argument, that in the case of light measurements in general we have to do with two stochastic processes X and Y, participating in the composition of a color. If ever in need of a statistical evaluation of the parameters a, b, c of the density from equation (15) above, we have at our disposal the maximum information entropy principle, for instance, giving their values by

$$a = \frac{\text{var}(y)}{D}, \; c = \frac{\text{var}(x)}{D}, b = -\frac{\text{cov}(xy)}{D},$$
$$D \equiv \text{var}(x)\text{var}(y) - [\text{cov}(xy)]^2. \qquad (6.15)$$

Here "*var*" and "*cov*" denote the variance and the covariance of the experimental data on X and Y.

This characterization of the color measurements the so-called dichromatic characterization is closely related to a plane centric affine geometry. This is to say that if one insists in characterizing the measurements of light in a plane, which is obviously the natural way to consider these measurements ever since the first ideas on light came out [20], the geometry of this plane is the centric affine geometry. The group of this plane geometry is given by the infinitesimal generators

$$X_1 = y\frac{\partial}{\partial x}, X_2 = \frac{1}{2}(x\frac{\partial}{\partial x} - y\frac{\partial}{\partial y}), \;\; X_3 = -x\frac{\partial}{\partial y}, \qquad (6.16)$$

while the group of the space of values a, b, c is given by infinitesimal generators

$$X_1 = -a\frac{\partial}{\partial b} - 2b\frac{\partial}{\partial c}, \; X_2 = -a\frac{\partial}{\partial a} + c\frac{\partial}{\partial c}, \;\; X_3 = 2b\frac{\partial}{\partial a} + c\frac{\partial}{\partial b}. \qquad (6.17)$$

These are two realizations of the same $SL(2, R)$ algebraical structure. The second one has intransitive action, which allows transitivity only along specific manifolds, given by constant discriminant of the quadratic form from the exponent of equation (6.14).

The probability density (6.14) itself can be presented as a joint invariant of the two actions (6.16) and (6.17), with the help of Stoka theorem [21]. According to this theorem, any joint invariant of the two actions is an arbitrary continuous function of the two algebraic formations

$$ax^2 + 2bxy + cy^2, \quad ac - b^2. \tag{6.18}$$

Obviously (6.14) is only a special case of this theorem. By the same token, the straight lines through origin $x = y = 0$ can be presented as joint invariants of two actions (6.16), while the joint invariants of two actions (6.17), one in the variables a, b, c, the other in the variables α, β, γ, say, are arbitrary functions of the following three algebraic formations [22]:

$$\alpha\gamma - \beta^2, \; ac - b^2, \; a\gamma + c\alpha - 2b\beta. \tag{6.19}$$

These facts can give good reasons for a few further observations related to the classical theory of colors.

The argument along these lines allows us to put forward, both in a classical Hooke-type theory, and in the modern theory of fluctuations of light, the Newtonian idea of a general two-dimensional symmetry involved in the description of light and matter. This leads further to a representation of color in connection with MacAdam discovery of the meaning of quadratic forms for which the discriminant from equation (6.18) is positive [23]. First of all, we have to be a little more specific about the probability densities like that from (6.14). Thus, for instance, consider that the background light color in a measurement process on a certain plane has the normal density

$$p_{XY}(x, y|\alpha, \beta\gamma) \equiv \frac{\sqrt{\alpha\gamma - \beta^2}}{2\pi} \exp\{-\frac{1}{2}(\alpha x^2 + 2\beta xy + \gamma y^2)\}, \tag{6.20}$$

in two variables X and Y, of which we don't know too much for now, other than that they are characterized by the statistics α, β and γ, as suggested before. All we know for sure is that, in practice, X and Y are some kind of projections on unspecified planes, that happen to be experimentally realizable, and that they represent two colors (the so-called property of dichromacy). At this moment, the theory is therefore dichromatic. Now, let us say that the two processes are jointly participating to give a third process, and all we know of this participation is that it is some kind of addition of them.

More specifically, we will suppose that this third process is a kind of weighted sum of the two processes, having the general form

$$Z = \mu X + \nu Y. \tag{6.21}$$

This is, for instance, the case of initial conditions in the case of the harmonic oscillator, under the condition of a proper gauging of light. The participations and are, in this particular case, given by the two solutions of a second order differential equation. The problem now is to find the probability density of the stochastic process Z. This can be done by following a known statistical routine, and the final result is

$$P_Z(z) = \sqrt{\frac{\alpha\gamma - \beta^2}{2\pi(\alpha\nu^2 - 2\beta\mu\nu + \gamma\mu^2)}} \exp\{-\frac{1}{2}\frac{\alpha\gamma - \beta^2}{2\alpha\nu^2 - 2\beta\mu\nu + \gamma\mu^2}z^2\}. \tag{6.22}$$

This is a Gaussian type probability density, having a zero mean and the variance

$$\sigma_Z^2 = \frac{\alpha\nu^2 - 2\beta\mu\nu + \gamma\mu^2}{\alpha\gamma - \beta^2}. \tag{6.23}$$

Such a probability density is particularly attractive in constructing the one related to characterizing the differentials of the three statistics, given their values.

Indeed, the equation (6.23) is indication of the nature of an "intensity variable" so to speak. It obviously satisfies the Stoka theorem, and indicates that the quadratics are essential in the statistics related to the "trichromacy" theory of colors. One can see directly that the trichromacy is due to the fact that there is a "dichromatic" moment in the theory of color space, related to the experimental procedures. Indeed, as we mentioned before, from algebraical point of view, the set of binary quadratics like those occuring in the exponent of a bivariate Gaussian, is a linear three-dimensional space. Whence the basic theoretical support for the idea that the color space should be three-dimensional, even though not necessarily Euclidean. This, of course, gives even more reasons for considering the quadratic as fundamental in the theory of light colors.

There should be, therefore, a way to the color of light, giving consistency to the ideas regarding the trichromacy of light colors directly through a general quadratic statistical variable $Z(X, Y)$, obtained, by dichromacy, in the measurement process of its values:

$$z(x, y) \equiv \frac{1}{2}(ax^2 + 2bxy + cy^2). \tag{6.24}$$

This statistical variable then characterizes a specific plane of illumination, no matter of the orientation of that plane, because the quadratic is form-invariant by any central projection. We have thus to find the probability density of this variable, under condition that the plane of light is characterized by the a priori probability density as given, for instance, in equation (6.20). That probability density satisfies, of course, the Stoka theorem, and the probability density of Z should also satisfy that theorem, in the precise sense that it must be a function of the algebraical formations from equation (6.19). This leaves us with a functionally undetermined probability density though, even if we impose some natural constraints in order to construct it.

Proceeding nevertheless directly, in the usual manner of the statistical practice, we are able to solve the problem, at least in this particular case. Thus, we have to find first the characteristic function of the variable (6.24). As known, this is the expectation of the imaginary exponential of Z, using (6.20) as probability density. Performing this operation directly, we get:

$$\langle e^{i\xi Z}\rangle = \frac{1}{2\pi\sqrt{1+(i\xi)\frac{a\gamma+c\alpha-2b\beta}{ac-b^2}+(i\xi)^2\frac{a\gamma-\beta^2}{ac-b^2}}}. \tag{6.25}$$

In view of (6.19), this characteristic function certainly satisfies the Stoka theorem, which thus reveals its right place in the physical theory. Like the Wien displacement law in the case of selection of the physically correct spectrum for blackbody radiation, the Stoka theorem should also serve for the selection of the right probability density in the case of light colors in general. Anyway, the sought for probability density can then be found by a routine Fourier inversion of (6.25), based on tabulated formulas [24]:

$$p_z(z|a,b,c) = \sqrt{AB}\exp\left(-\frac{A+B}{2}z\right)\cdot I_0\left(\frac{A-B}{2}z\right). \tag{6.26}$$

Here I_0 is the modified Bessel function of order zero, and A, B are two constants to be calculated from the formulas

$$A+B = \frac{2b\beta - a\gamma - c\alpha}{ac-b^2},\quad AB = \frac{a\gamma-\beta^2}{ac-b^2},\quad A > b. \tag{6.27}$$

Again, this probability density obviously satisfies the Stoka theorem, as it is a function of the joint invariants from equation (6.19). And so do the mean and the standard deviation of the variable Z, for they can be calculated as

$$\langle Z\rangle \equiv \frac{1}{2}\frac{A+B}{AB} = \frac{1}{2}\frac{2b\beta - a\gamma - c\alpha}{ac-b^2}, \tag{6.28}$$
$$var(Z) \equiv \frac{1}{2}\frac{A^2+B^2}{A^2B^2} = \frac{1}{2}\left(\frac{2b\beta - a\gamma - c\alpha}{ac-b^2}\right)^2 - \frac{\alpha\gamma-\beta^2}{ac-b^2}.$$

We thus have the interesting conclusion that the essential statistics related to variable Z do not depend but on the coefficients of the background color distribution, and the values of the parameters entering the expression of the color Z. On one hand, this means that the geometry of the color space is dictated by the statistical characteristics of the plane of projection and by the physics describing the color, naturally incorporated in the variable Z. For instance Z can represent the energy of a harmonic oscillator, or even the wavelength of light when described by the wave surface. On the other hand, our result shows that the color space is actually characterized by a Riemannian metric of negative curvature, which is the current tenet in the theory of color. Let"s show this.

6.6 Light as a Stochastic Process

One usually insists, and with good reasons at that, upon the fact that the geometry of the color space is not an Euclidean one, but a general Riemannian geometry (see [16,25,26]). In such circumstances, the Riemannian metric carries a special statistical significance whereby the components of the metric tensor are covariances of the three color coordinates [27, 28]. However, this meaning of the metric does not seem to be theoretically secured. Yet one works this way, and the results confirm the manner of approach everywhere in the classical theory of color. There should be therefore some fundamental truth there, whose formal expression is not yet obvious. And there is, of course, a fundamental truth here, giving deeper physical grounds to the holographic principle in an unexpected form related to the classical theory of colors.

First, the previous statistical theory can help us secure, from a theoretical point of view, a purely statistical connotation in the color space. Assume indeed, that a, b and c are some variations of the "background" parameters α, β and γ, respectively. It thus turns out that this variation, dZ say, of the color measure Z, is dictated only by the variations of its coefficients, and it is a process having, according to equation (6.28), the following expectation and variance:

$$\overline{dZ} \equiv \frac{1}{2}\frac{A+B}{AB} = \frac{1}{2}\frac{2\beta d\beta - \alpha d\gamma}{\alpha\gamma - \beta^2}, \tag{6.29}$$

$$\overline{[\Delta(dZ)]^2} \equiv \frac{1}{2}\frac{A^2+B^2}{A^2B^2} = \frac{1}{2}\left(\frac{2\beta d\beta - \gamma d\alpha - \alpha d\gamma}{\alpha\gamma - \beta^2}\right)^2 - \frac{d\alpha d\gamma - (d\beta)^2}{\alpha\gamma - \beta^2}.$$

Here a bar over the symbol means average using the probability density given by equation (6.26). From these formulas we get a statistic having a special geometrical meaning:

$$\begin{aligned} &\overline{[\Delta(dZ)]^2} - \overline{dZ} \equiv \overline{dZ^2} - \overline{2dZ}^2 \\ &= \frac{1}{4}\left(\frac{2\beta d\beta - \gamma d\alpha - \alpha d\gamma}{\alpha\gamma - \beta^2}\right)^2 - \frac{d\alpha d\gamma - (d\beta)^2}{\alpha\gamma - \beta^2}. \end{aligned} \tag{6.30}$$

The right hand side of this formula carries indeed a special meaning: it is the Riemannian metric which can be built by the methods of absolute geometry for the space of the 2×2 matrices, having the singular matrices as points of the absolute quadric [29]. In fact, one can prove, and we will show this immediately, that the quadratic form (6.30) is just the Cartan-Killing metric of the certain action of the 2×2 real matrices. This is indeed of the quadratic form

$$\frac{1}{4}(\omega_2^2 - 4\omega_1\omega_3), \tag{6.31}$$

where $\omega_{1,2,3}$ are three 1-forms representing three conservation laws of the $SL(2, R)$, and has the exquisite interpretation already mentioned. Meanwhile, let"s notice that, from a stochastic point of view, the process of physical variation of the parameters of the quadratic form is “almost” a Lvy-type process with three parameters [30], in the sense that the elementary distance is decided by the variance function. This validates indeed the statistical interpretation of the metric of the space of colors, but raises instead another problem related to the coordinates representing the colors. This problem indicates, in turn, the feasibility of another, more special, approach of the geometry of colors, leading to the idea of Yang-Mills fields even in the classical case.

6.7 Resnikoff's Special Theory of Colors

Notice indeed that, actually, it is not the variable dZ we are after, but the parameters $d\alpha, d\beta$ and $d\gamma$, and they can be assumed to have zero averages, without any problem. Equations (6.29) and (6.30) are then just constraining control equations, related to a space coherence of light for instance. Indeed, we usually measure the wavelength in order to get the characteristics of light and, when referred to the wave surface, the wavelength is a quadratic form in the parameters of the plane of dichromatic measurements. Howard Resnikoff introduced as representative for what he calls the "perceptual lights" a set of 2×2 symmetric matrices [31]:

$$\alpha \equiv \begin{pmatrix} \alpha & \beta \\ \beta & \gamma \end{pmatrix} \tag{6.32}$$

involving directly the second-order statistics. The determinant of this symmetric matrix is taken as the brightness variable of the light, to be constructed from the three basic color perceptions. Resnikoff suggests that the entries of the matrix (6.32) are to be taken as color coordinates. In that case the coordinate can be chosen to be the regular "B" the quantifier for the "blue" color in an RGB color scheme of course, in the cases where the brightness of light thus calculated is positive. For a certain situation β has therefore to play the part of a correlation when statistically considered in the case of dichromatic basic variables. The choice is not unique, for there are three manners of calculating this brightness on a certain range of the color parameters RGB, in order to satisfy the positivity requirement, but let us go with it just for the sake of illustration. Thus, if we take, in the manner of Resnikoff:

$$\xi = \sqrt{\alpha\gamma - \beta^2}, \quad u = \frac{\beta}{\alpha}, \quad v = \frac{\alpha\gamma - \beta^2}{\alpha} \tag{6.33}$$

the matrix (6.32) becomes

$$\alpha = (\xi/v) \begin{pmatrix} 1 & u \\ u & u^2 + v^2 \end{pmatrix}. \tag{6.34}$$

In this case we have by direct calculation:

$$\alpha^{-1} d\alpha = d\ln(\xi/v) \begin{pmatrix} 1 & 0 \\ 0 & 1 \end{pmatrix} + (1/v^2) \begin{pmatrix} -udu & -(u^2 - v^2)du - 2uvdv \\ du & udu + 2vdv \end{pmatrix} \tag{6.35}$$

and the Resnikoff metric is just the Cartan-Killing metric of this group of matrices, given by:

$$tr[(\alpha^{-1}d\alpha)\cdot(\alpha^{-1}d\alpha)] = 2\left\{\left(\frac{d\xi}{\xi}\right)^2 + \frac{du^2+dv^2}{v^2}\right\}. \qquad (6.36)$$

Now, the matrix (36) has the general form:

$$\alpha^{-1}d\alpha = d(\ln x)\begin{pmatrix}1 & 0\\ 0 & 1\end{pmatrix} + (1/v^2)\begin{pmatrix}-udu-vdv & -(u^2-v^2)du-2uvdv\\ du & udu+vdv\end{pmatrix} \qquad (6.37)$$

and carries a special meaning in the geometrical theory of color. In order to reveal this meaning let's consider the quadratic forms in their utmost generality, from the general standpoint that their coefficients do represent lights or color coordinates, as suggested by Resnikoff.

6.8 Differential Dichromacy: the MacAdam Ellipses

The general equation of a conic section is a quadratic equation of the form

$$f(x,y) \equiv \alpha x^2 + 2\beta xy + \gamma y^2 + \gamma y^2 + 2ax + 2by + c = 0. \qquad (6.38)$$

This time in the quadratic form we have included the possibility of an arbitrary center not just the origin whose coordinates are related to the coefficients a, b through a linear homogeneous relation determined by α, β and γ. There is a merit, given by handling simplicity among others, in using again the "notation of Dirac". This also allows for a suggestive interpretation of the final geometrical results. In such notation the equation (6.38) can be written as

$$f(x,y) \equiv \langle x|\boldsymbol{\alpha}|x\rangle + 2\langle a|x\rangle + c = 0, \qquad (6.39)$$

where we used the following identification:

$$|a\rangle \equiv \begin{pmatrix}a\\ b\end{pmatrix} \therefore \langle a| \equiv \begin{pmatrix}a\\ b\end{pmatrix}^t = (ab).$$

This vector represents the relative position of the center of the conic in the known geometrical sense:

$$\boldsymbol{\alpha}|x_c\rangle + |a\rangle = |a\rangle \therefore |x_c\rangle = -\boldsymbol{\alpha}^{-1}|a\rangle. \tag{6.40}$$

If we refer the conic to this center, by means of the translation

$$\begin{pmatrix} x \\ y \end{pmatrix} = \begin{pmatrix} x_c \\ y_c \end{pmatrix} + \begin{pmatrix} \xi \\ \eta \end{pmatrix} \leftrightarrow |x\rangle = |x_c\rangle + |\xi\rangle$$

the equation (40) becomes purely quadratic in coordinates, although otherwise inhomogeneous:

$$\langle \xi|\boldsymbol{\alpha}|\xi\rangle - \langle x_c|\boldsymbol{\alpha}|x_c\rangle + c = 0. \tag{6.41}$$

This algebra is now used in constructing an argument for the matrix representation of colors.

Within the framework of Resnikoff representation presented above, the problem of identification of a center of color in a plane of measurement what we would like to call the MacAdam"s problem [23] – has an explicit algebraical expression. Indeed, we can simply represent the repeated targeting of "the same geometrical color center by the differential equations $dx_c = dy_c = 0$. Then the condition (6.40) comes formally down to the following matrix differential equation:

$$|a\rangle \equiv d(\boldsymbol{\alpha}^{-1}|a\rangle) = (d\boldsymbol{\alpha}^{-1})|a\rangle + \boldsymbol{\alpha}^{-1}|da\rangle. \tag{6.42}$$

Obviously this equation limits the set of possible conics having the same geometric center. Using the definition of the inverse of a matrix, to the effect that $\boldsymbol{\alpha}^{-1} \cdot \boldsymbol{\alpha}$ is the identity matrix, one can easily prove by direct differentiation the matrix differential relation $d\boldsymbol{\alpha}^{-1} = -\boldsymbol{\alpha}^{-1} \cdot d\boldsymbol{\alpha} \cdot \boldsymbol{\alpha}^{-1}$, so that from equation (6.40) we must have

$$|da\rangle = (d\boldsymbol{\alpha} \cdot \boldsymbol{\alpha})|a\rangle. \tag{6.43}$$

Thus the condition of fixed center comes actually down to a certain evolution of the vector $|a\rangle$, dictated by the matrix of the quadratic form from the equation of the conic section and its variation. In detail, the equation (6.43) can be written as

$$\begin{pmatrix} da \\ db \end{pmatrix} = \frac{1}{\alpha\gamma - \beta^2} \begin{pmatrix} \gamma d\alpha - \beta d\beta & \alpha d\beta - \beta d\alpha \\ \gamma d\beta - \beta d\gamma & \alpha d\gamma - \beta d\beta \end{pmatrix} \begin{pmatrix} a \\ b \end{pmatrix}. \tag{6.44}$$

The matrix governing the evolution in the right hand side of this equation can be further adjusted to a special form:

$$\boldsymbol{\Omega} \equiv (\ln\sqrt{\Delta})\begin{pmatrix}1 & 0\\ 0 & 1\end{pmatrix} + \begin{pmatrix}-\omega_2/2 & \omega_1\\ \omega_3 & \omega_2/2\end{pmatrix}, \quad \Delta = \alpha\gamma - \beta^2, \qquad (6.45)$$

Δ is therefore the determinant of , i.e. Resnikoff"s brightness squared, and we denoted

$$\omega_1 = \frac{\alpha d\beta - \beta d\alpha}{\Delta}, \quad \omega_2 = \frac{\alpha d\gamma - \gamma d\alpha}{\Delta}, \quad \omega_3 = \frac{\beta d\gamma - \gamma d\beta}{\Delta}, \qquad (6.46)$$

three differential forms generated by the elements of the matrix of quadratic form representing the family of local colors, and their differentials. When calculated in the coordinates from equation (6.33) these differential forms are

$$\omega_1 = \frac{du}{v^2}, \quad \omega_2 = 2\frac{udu + vdv}{v^2}, \quad \omega_3 = \frac{(u^2 - v^2)du + 2uvdv}{v^2}, \qquad (6.47)$$

showing explicitly that the matrix from equation (6.45) is the transposed of that from equation (6.37).

Thus, the proposed representation of Resnikoff"s has actually a firm physical basis, in relation to MacAdam's ellipses. Indeed, assume that we are to identify a certain center, as in MacAdam experiments. The center is the one position satisfying equation (6.40), and therefore asks for the differential correlation (6.43), which turns out to be an equation of motion for the vector $|a\rangle$. When the Resnikoff's matrix is taken as shown, i.e. representing an ellipse, then the motion of the center $|a\rangle$ itself is along an ellipse, which is the real case with MacAdam results. Therefore, the MacAdam's ellipse gives indeed a statistical interpretation to differentials of the elements of color in Resnikoff's representation.

6.9 A General Dynamics of Color

Now, a few algebraical relations among the differential forms (6.47) are in order. They form a basis of a $SL(2,R)$ algebra. The following differential relations can be directly calculated:

$$d \wedge \omega_1 = \frac{\alpha}{\sqrt{\Delta}}\Theta d \wedge \omega_2 = \frac{2\beta}{\sqrt{\Delta}}\Theta; d \wedge \omega_3 = \frac{\gamma}{\sqrt{\Delta}}\Theta. \qquad (6.48)$$

where Θ is the differential 2-form

$$\Theta \equiv \frac{\alpha d\beta \wedge d\gamma + \beta d\gamma \wedge d\alpha + \gamma d\alpha \wedge d\beta}{\Delta^{3/2}}. \tag{6.49}$$

The 2-form Θ is closed because it is the exterior differential of a 1-form:

$$\Theta \equiv d \wedge \psi, \qquad \equiv \frac{\alpha+\gamma}{\sqrt{\Delta}} d\left(tan^{-1}\frac{2\beta}{\alpha-\gamma}\right) \tag{6.50}$$

representing the Hannay angle of this problem. In our context it gives a way to "objectify", so to speak, the subjective experimental evaluations of colors, and has certainly everything in common with the original angle [32, 33].

On the other hand, we can verify the following relations:

$$\omega_1 \wedge \omega_2 = \frac{\alpha}{\sqrt{\Delta}}\Theta, \;\; \omega_2 \wedge \omega_3 = \frac{\gamma}{\sqrt{\Delta}}\Theta, \;\; \omega_3 \wedge \omega_1 = -\frac{\beta}{\sqrt{\Delta}}\Theta. \tag{6.51}$$

Thus, from (6.48) and (6.51) we have the characteristic equations of a $sl(2, R)$ structure:

$$d \wedge \omega_1 - \omega_1 \wedge \omega_2 = 0, \;\; d \wedge \omega_3 - \omega_2 \wedge \omega_3 = 0, \;\; d \wedge \omega_2 + 2(\omega_3 \wedge \omega_1) = 0. \tag{6.52}$$

Using these relations we can draw an important conclusion: the quadratic forms associated with the matrix in Resnikoff representation of light perceptuals are actually fluxes of color in the color space, induced by the "subjective" uncertainty in determining a color. Indeed, the quadratic form conserved along MacAdam"s evolution can be written as $\langle a|\omega|a\rangle$, where ω is a symmetric matrix of 1-forms in Resnikoff's perceptuals. One can thus construct the 2-form

$$\langle a|d \wedge \omega|a\rangle = \langle a|\boldsymbol{\alpha}|a\rangle \frac{\Theta}{\sqrt{\Delta}}, \quad \boldsymbol{\omega} \equiv \begin{pmatrix} \omega_1 & \omega_2/2 \\ \omega_2/2 & \omega_3 \end{pmatrix}, \tag{6.53}$$

where we have used the equations (6.48). As the 2-form Θ is a flux, the analogous of the solid angle in the usual Euclidean space, the quadratic form $\langle a|\alpha|a\rangle$ is indeed the intensity of a flux of colors in the color space thus defined. One might say that the human eye is driven, in evaluating the light, by a flux of colors correlated to Hannay"s angle.

6.10 Perspectives

The concepts, especially the physical ones have their internal dynamics. The present work advocates the idea of a continuity of this dynamics: first of the quantum theory with respect to classical theory, then of both theories as regarded through the modern idea of the holographic principle. Resuming this last principle, one could say that the light is a universal model of the physical world. We just tried to make this statement more explicit.

In order to conclude the work nothing would come better than a few words excerpted from the articles that founded the holographic principle in its modern form. First, the words of Gerardus "t Hooft: We would like to advocate here a somewhat extreme point of view. We suspect that there simply are not more degrees of freedom to talk about than the ones one can draw on a surface The situation can be compared with a hologram of a three dimensional image on a two-dimensional surface. The image is somewhat blurred because of limitations of the hologram technique, but the blurring is small compared to the uncertainties produced by the usual quantum mechanical fluctuations. The details of the hologram on the surface itself are intricate and contain as much information as is allowed by the finiteness of the wavelength of light read the Planck length. ([8], our Italics). Involving the quantum mechanics here has raised problems. Leonard Susskind advocates no existing contradiction, and is even more precise as to the involvement of the quantum limit:

According to "t Hooft it must be possible to describe all phenomena within V by a set of degrees of freedom which reside on the surface bounding V. The number of degrees of freedom should be no larger than that of a two dimensional lattice with approximately one binary degree of freedom per Planck area. In other words the world is in a certain sense a two dimensional lattice of spins ([9], our Italics). We have shown that these statements do have a historical lineage, starting with Newton, and continuing with the quantum theory of light, which occasioned the idea of holography in the first place. The present work showed that Newton"s theory of light means actually that the light supports the idea of an abstract symmetry related to color. The corollary of Newton"s work can be properly understood by referring it to Hooke"s rational theory of colors. In short it states: is not the geometrical form of the light ray that should prevail, but the general two-dimensional symmetry. The Planck"s theory of quanta points out to the very same general symmetry property of physical fields. No wonder then, the holography, as well as its quantal basis, should have roots in the classical theory of light, and the holographic principle thus turns out to be a physically sound universal principle, inasmuch as the light carries the information in the universe. But there is more to it.

First, the light can be taken as a sound physical model of the theory of interactions of material particles, defined in the modern way, i.e. experimentally, which is plainly a Newtonian way of seeing the particles. The particles here are not material points in the classical sense, but have a space extension. Thus they have a surface, and this surface is the one through which the interaction takes place. The present work shows that the interaction is then described by a $SL(2, R)$ Lie algebra. This approach offers a rationale to the classical theory of color, seen as a theory of interaction of light with the matter. Moreover, it offers a general view of the theory of fundamental interactions by what we call a Resnikoff-type of representation of interactions.

But the implications of a theory that uses a Resnikoff's representation of colors, whereby they are quantitatively described by the entries of a 2×2 symmetric real matrix, are far more intricate from physical theoretical point of view. Indeed, such a representation has an outstanding theoretical meaning. A matrix is obviously an element of a noncommutative algebra, which can be simply a Yang-Mills field. It turns out that this classical theory of colors is plainly a Yang-Mills theory. It completes the classical theory of light in a natural way, by including the color in it. The classical electromagnetic theory, even though undoubtedly a gauge theory, is not a Yang-Mills theory yet. The present work shows that it takes considerations of color of light in order to render to the theory of light a plain Yang-Mills character. The modern "technicolor" for instance, should be a genuine classical concept. From this point of view, the light itself actually enters the realm of quantum chromodynamics, as it should naturally do, for the everyday color is related to light. But there is more to it: if the mechanism of color is the one explaining the strong interactions, then this color should be classical too. Thus one might figure out why the noncommutativity is the essential ingredient allowing asymptotic freedom in the case of strong interactions: after all, the light is a model of interaction everywhere in the universe, at any level! For more details see [34].

References

1. Hobbes, T.: Tractatus Opticus, in Thomae Hobbes Malmesburiensis Opera Philosophica Quae Latine Scripsit Omnia, J. Bohn, London, 5, 215, (1839).
2. Levi-Civita, T.: The Absolute Differential Calculus, Blackie & Son Limited, London and Glasgow, (1927).
3. Hooke, R.: Micrographia, or Some Physiological Descriptions of Minute Bodies Made by Magnifying Glasses, Martyn & Allestry, London (1665).
4. Hooke, R.: The Posthumous Works of Robert Hooke, Johnson Reprint Corporation, New York, (1969).
5. Shapiro, A. E.: Kinematic Optics: A Study of the Wave Theory of Light in the Seventeenth Century, Archives for the History of Exact Sciences, 11, 134, (1973).
6. Shapiro, A. E.: Newton"s Definition of a Light Ray and the Diffusion Theories of Chromatic Dispersion. Isis., 66, 194, (1975).
7. Newton, Sir Isaac: Opticks, or a Treatise of the Reflections, Refractions, Inflections & Colours of Light. Dover Publications, Inc., New York, (1952).
8. "t Hooft, G.: Dimensional Reduction in Quantum Gravity, arxiv: gr-qc/9310026, (1993).
9. Susskind, L.: The World as a Hologram, arxiv: hep-th/9409089, (1994).
10. Fresnel, A.: Mémoire sur la Double Réfraction. Mmoirs de l'Académie des Sciences de l'Institute de France, 7, 45, (1827).
11. Finikov, S. P.: A Course on Differental Geometry, OGIZ, Moscow, (1952).
12. Flanders, H.: Differential Forms with Applications to the Physical Sciences, Dover Publications, New York, (1989).
13. Finikov, S. P.: Cartan's Method of Exterior Forms in Differential Geometry, OGIZ, Moscow, (1948).
14. Guggenheimer, H. W.: Differential Geometry, Dover Publications, New York, (1977).

15. Lowe, P. G.: A Note on Surface Geometry with Special Reference to Twist. Mathematical Proceedings of the Cambridge Philosophical Society, 87, 481, (1980).

16. Schrödinger, E.: Grundlinien einer Theorie der Farbenmetrik im Tagessehen I, II, III, Annalen der Physik., 63, 397, 427, 481, (1920).

17. Planck, M: Original Papers in Quantum Physics (1900), translated by D. Ter-Haar and S. G. Brush, and Annotated by H. Kangro, Wiley & Sons, New York, (1972).

18. Mazilu, N.: Bulletin of the Polytehnic Institute of Iaşi. 56 (2010), 69; A Case Against the First Quantization, viXra.org/quantu physics/1009.0005/.

19. MacAdam, D. L: Journal of the Optical Society of America, 67, 839, (1977).

20. Hoffman, W. C.: The Lie Algebra of Visual Perception. Journal of Mathematical Psychology, 3, 65, (1966).

21. Stoka, M. I.: Géométrie Intégrale, Mémorial des Sciences Mathémati Gauthier-Villars, Paris, (1968).

22. Mazilu, N.: Supplemento ai Rendiconti del Circolo Matematico di Palermo, 77, 415, (2004).

23. MacAdam, D. L: Journal of the Optical Society of America, 32, 247, (1942).

24. Gradshteyn, I. S., Ryzhik, I. M.: Table of Integrals, Series and Products, Seventh Edition, A. Jeffrey & D. Zwillinger Editors. Academic Press-Elsevier, Waltham, (2007).

25. MacAdam, D. L: Sources of Color Science. The MIT Press, Cambridge, MA & London, UK, (1970).

26. Wyszecki, G., Stiles, W. S.: Color Science: Concepts and Methods, Quantitative Data and Formulae, R. E. John Wiley & Sons, New York, (1982).

27. Silberstein, L.: Journal of the Optical Society of America, 28, 63, (1938).

28. Silberstein, L.: Journal of the Optical Society of America, 33, 1, (1943).

29. Mazilu, N., Agop, M.: Skyrmions-a Great Finishing Touch to Classical Newtonian Philosophy, Nova Publishers, New York, (2012).

30. Lévy, P.: Processus Stochastiques et Mouvement Brownien. Gauthier-Villars, Paris, (1965).

31. Resnikoff, H. L.: Journal of Mathematical Biology, 1, 97, (1974).

32. Hannay, J. H.: Journal of Physics A: Mathematical and General, 18, 221, (1985).

33. Berry, M. V.: Journal of Physics A: Mathematical and General, 18, 15, (1985).

34. Mazilu, N., Agop, M., Gaţu, I., Iacob, D.D., Butuc, I., Ghizdovăţ, V.: The classical theory of light colors: a paradigm for description of particle interactions, International Journal of Theoretical Physics, (2016).

